W9-BQS-299

Graphic Discovery

Graphic Discovery

A Trout in the Milk and Other Visual Adventures

HOWARD WAINER

Princeton University Press Princeton and Oxford

Published by Princeton University Press,
41 William Street, Princeton, NJ 08540
In the United Kingdom: Princeton University Press,
3 Market Place, Woodstock, Oxfordshire OX20 ISY

Library of Congress Cataloging-in-Publication Data

Wainer, Howard.
Graphic Discovery: a trout in the milk and other visual adventures / Howard Wainer.
p. cm
Includes bibliographical references and indexes.
ISBN 0-691-10301-1 (cloth : acid-free paper)
1. Mathematical statistics—Graphic methods.
2. Playfair, William, 1759–1823.
3. Mathematical statistics—History—18th century. I. Title.
QA276.3.W29 2005
519.5—dc22 2003066083

British Library Cataloging-in-Publication Data is available

This book has been composed in
Adobe Garamond and Akzidenz Grotesk
Printed on acid-free paper. ∞
pup.princeton.edu
Printed in the United States of America
10 9 8 7 6 5 4 3 2 1

Frontispiece: A portrait of Sir Francis Galton (1822–1911) by Susan Slyman showing him among some of his graphic inventions. In the upper left corner is the graph of his discovery of the anticyclonic movement of air around low-pressure zones that forms the basis of chapter 8.

To William Playfair and John Wilder Tukey,

revolutionaries both,

without whom this book

would have been unimaginable

Contents

Preface

At dinner one night, Winston Churchill is said to have turned up his nose at the dessert, proclaiming, "This pudding has no theme." This comment gives voice to what was my greatest fear in writing this book. I knew before I began that although the general theme of this book would be the history and future of methods for the visual communication of information, it would be such a potpourri of graphical topics that the unwary reader might see my pudding the same way that Sir Winston saw his.

My tale begins with the eighteenth-century origins of the art of data display, when its logic and methods emerged, full-grown, from the mind of William Playfair out of the ashes of seventeenth-century Cartesian rationalism. This remarkable Scot was, at first acquaintance, an unlikely source for such a marvelous invention, but after digesting the facts you will be hard pressed to imagine any one else having been able to do it. When, finally, the story ends, we find ourselves perched high atop the shoulders of John Wilder Tukey, peering into the misty future and trying to divine what graphical tools will best serve the needs of the data-rich twenty-first century. In the sandwich between these two giants is stuffed a mixture of graphical investigations, each providing a taste of the illumination that is possible when the tools are used correctly. And, when they are misused, I show how the vast darkness of the topic under investigation can be effectively obscured.

There are a number of contemporary books on the topic of the visual communication of information, some of them very wonderful. Edward Tufte's three books are the most wonderful of these. They are nontechnical in orientation and focus on the general issues of visual design. Bill Cleveland's and Lee Wilkinson's books provide an ample discussion of the technical issues in information design. Stephen Kosslyn has examined the relationships between cognitive science and visual display. With these fine resources available, there needs to be an important reason to put pixel to screen and prepare yet another one. I hope that after you have read *Graphic Discovery* you will agree with me that it was worth doing; that no one else has stirred together this mixture of historical and future graphical practice with a *nuage* of modern statistics to produce a visual pudding whose breadth of theme is more than made up for by its rich texture and its savory tang.

Modern scholarship is rarely a solitary activity, even when it seems to be. Authoring a book may seem a lonely task to an outside observer who sees only the back of someone leaning over a keyboard. But this is an incorrect impression. In the course of writing this book I was helped and abetted by an enormous cast of characters. It is my pleasure now to be able to express my gratitude, while exculpating them from responsibility for those flaws that remain.

I begin with a large thank-you to my employers during the time this book was conceived and written. First, to the Educational Testing Service, which was my professional home for twenty-one years, during which I learned much of what I now know about statistical graphics. Although little of my job explicitly had to do with the design of instruments to transmit information, I was permitted the time to indulge myself in this pursuit. Second, to my current employer, the National Board of Medical Examiners, which very generously allowed me the time to write this book and seemed to take pleasure in the way it was turning out.

This book would never have been completed without the help of Editha Chase, who has been my right hand. Among numerous other tasks that she performed cheerfully, Editha chased down the information I required for a number of the more obscure characters in this book and so provided the grist from which I milled the biographies on pages 151–72. She also kept everything straight.

At the National Board, a number of my colleagues graciously agreed to read and comment on various drafts as they were completed. Prominent among them are Steve Clyman, Bob Galbraith, and Don Melnick. The enthusiasm that Don showed for the title of chapter 13 led to my choice of the subtitle of the book.

Little that I do in graphics is independent of Edward Tufte, whose views on graphics helped shape my own and whose fine aesthetic sense, exemplified in his marvelous books, provides a model for all of us. Edward has generously allowed me to reprint a number of Playfair's figures that he reproduced so beautifully in the second edition of his legendary *Visual Display of Quantitative Information.* My gratitude is small compensation for his continuing help.

Several chapters are rewritten versions of earlier journal articles that either were done jointly or borrowed heavily from another's work. I am grateful for my coauthors' and colleagues' permission to include them here. Specifically, chapter 1 was drawn from work of Patricia Costigan-Eaves, Michael Macdonald-Ross, and Albert Biderman. Chapters 2 and 3 were drawn from articles that originally appeared in *Chance* magazine and were written principally by Ian Spence with a more minor contribution by me. Ian's scholarship and his fine prose are still evident in them. I am grateful to him for his permission to reprint them here as well as for his critical encouragement. Chapter 7 draws heavily from the scholarly investigations of Steve Ferguson, and chapter 8 was instigated by correspondence with Steve Stigler. Chapter 17 is drawn from a *Chance* article written by Catherine Njue, Sam Palmer, and me. Sam Palmer and Eric Bradlow collaborated with me in the preparation of the article that became chapter 22. As a ninth grader, Sam gathered the Princeton Cemetery data that eventually became figure 22.1.

Part III owes much to others. The biographical sketch of John Tukey that became chapter 19 was drawn principally from a more extensive biography that David Brillinger prepared. I also borrowed material from Fred Mosteller's biography that graces the introduction to each volume of Tukey's *Collected Works*. Last, I included information that Tukey's nephew Frank Anscombe provided that would not have been available anywhere else.

Chapters 20 and 21 grew from conversations I had with Tukey over the last year of his life. Almost without exception, whenever these conversations took place they were in the presence of Charles Lewis. Charlie helped to elicit further explication from Tukey when that was needed. He also helped me to understand John's sometimes Delphic pronouncements. Of course Charlie should not be held responsible for my rendering of them here. Earlier versions of this work appeared in *Chance* as well as in a chapter of the 2001 *Annual Review of Psychology* that I coauthored with Paul Velleman.

In addition to those mentioned above, this book has benefited from the sharp eyes and clear thinking of a number of colleagues. *Primus inter pares* is David Hoaglin, who covered every page with red ink. Although he deserves substantial credit for the clarity of what remains, he bears no responsibility for any errors that have sneaked in after he last saw the manuscript. I also received substantial help from Steve Stigler, who not only raised the issues discussed in chapter 8 but also provided invaluable aids for tracking down and understanding myriad historical details. Having him at the other end of my email has always been a comforting circumstance. I would also like to express my thanks to Jean-Paul Fournier and Steve Ferguson for their help in summarizing the life of Jacques Barbeau-Dubourg.

My friends and colleagues Tony Laduca, George Miller, Fred Mosteller, Malcolm Ree, David Thissen, and Lee Wilkinson all offered help and encouragement; for that I thank them. Special thanks to Princeton

University Press editor Vickie Kearn, whose enthusiasm for the project and critical judgment were more important than she can imagine. In addition, I am grateful to Madeleine Adams, Alison Kalett, Dimitri Karetnikov, Linny Schenck, and the rest of the production staff at Princeton University Press, whose professionalism and intolerance for error removed much of the spoor of sow's ear that characterized the initial manuscript.

Introduction

Let me begin with a few kind words about the bubonic plague. In 1538, Thomas Cromwell, the Earl of Essex (1485–1540),* issued an injunction (one of seventeen) in the name of Henry VIII that required the registration of all christenings and burials in every English parish. The London Company of Parish Clerks compiled weekly *Bills of Mortality* from such registers. This record of burials provided a way to monitor the incidence of plague within the city. Initially, these *Bills* were circulated only to government officials, principal among them the Lord Mayor and members of the King's Council.† They were first made available to the public in 1594, but were discontinued a year later with the abatement of the plague. However, in 1603, when the plague again struck London, their publication resumed on a regular basis.

The first serious analysis of the London *Bills* was done by John Graunt in 1662, but in 1710, Dr. John Arbuthnot, a physician to Queen Anne, published an article that used the christening data to support an argument (probably tongue-in-cheek) for the existence of God. These data also provide supporting evidence for the lack of existence of graphs at that time.

Figure 1 is a simple plot of the annual number of christenings in London from 1630 until 1710. The preparation of such a plot is straightforward, certainly requiring no more complex apparatus than was available to Dr. Arbuthnot in 1710. Moreover, as we will see in a moment, it is quite informative. Yet it is highly unlikely that Arbuthnot, or any of his contemporaries, ever made such a plot.

The overall pattern we see in figure 1 is a trend over these eighty years of an increasing number of christenings, almost doubling from 1630 to 1710. A number of fits and starts manifest themselves in substantial jiggles. Yet each jiggle, save one, can be explained. Some of these explanations are written on the plot. The big dip that began in 1642 can only partially be explained by the onset of the English Civil War. Surely the chaos common to civil war can explain the initial drop, but the war ended in 1649 with the beheading of Charles I at Whitehall, whereas the christenings did not return to their earlier levels until 1660.‡ Graunt offered a more complex explanation that involved the distinction between births and christenings, and the likelihood that Anglican priests would not enter children

* Thomas Cromwell was an English statesman who had a successful career as an administrator and advisor to the king. Among his other accomplishments, he arranged Henry VIII's divorce from Catherine of Aragon (and was largely responsible for the beheading of Sir Thomas More [1478–1535] in the process). Ironically, five years later, he was done in by Henry's aversion to Anne of Cleves, a consort of Cromwell's choosing, when he was consequently sent to the Tower of London and beheaded. Perhaps specifying the consequences of failure in this way would provide a workable pathway toward helping to improve the efficacy of computer dating services.

† This exposition is heavily indebted to the scholarly work of Sandy Zabell, to whose work the interested reader is referred for a much fuller description (Zabell 1976). It was Zabell who first uncovered Arbuthnot's clerical error.

‡ The year 1660 marked the end of the protectorate of Oliver Cromwell and the beginning of the restoration.

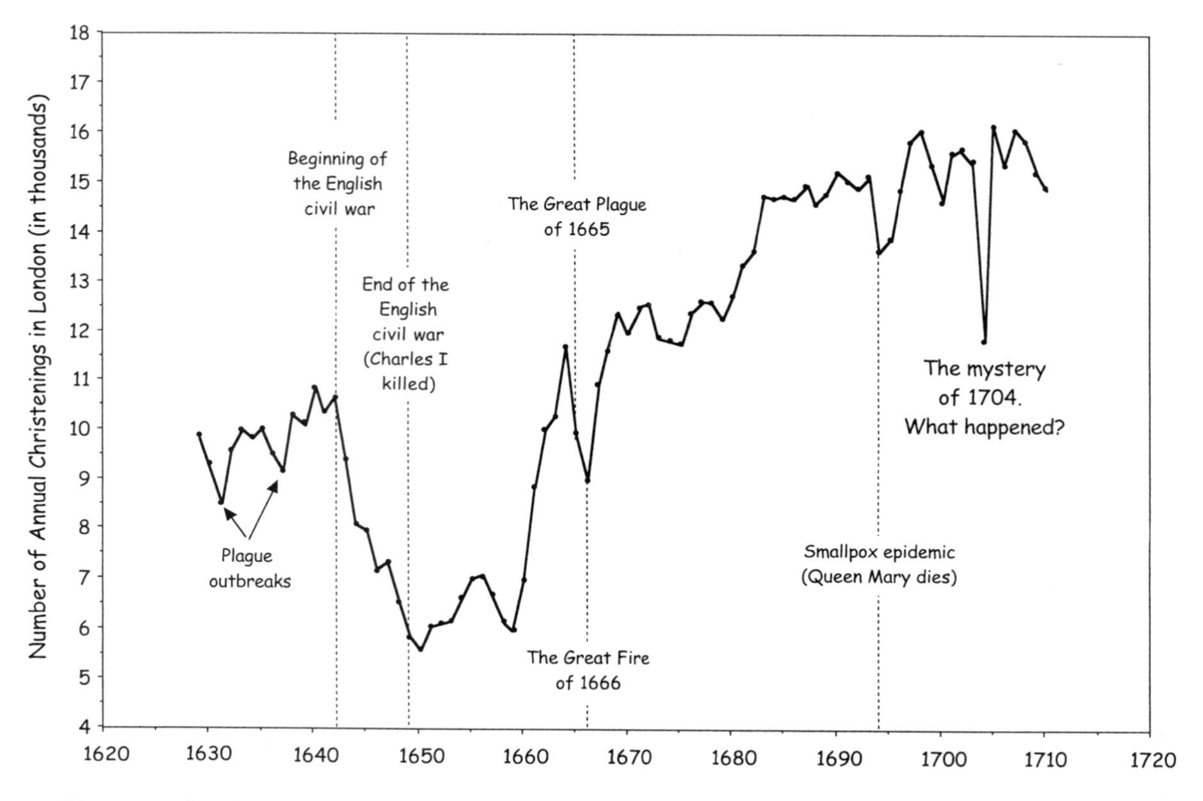

Figure 1. A plot of the annual christenings in London between 1630 and 1710 from the London *Bills of Mortality*. These data were taken from a table published by John Arbuthnot in 1710.

born to Catholics or Protestant dissenters into the register.

Many of the other irregularities observed are explained in figure 1, but what about the mysterious drop in 1704? That year has about four thousand fewer christenings than one might expect from observing the adjacent data points. What happened? There was no sudden outbreak of war or pestilence, no great civil uprising, nothing that could explain this enormous drop.

The plot not only reveals the anomaly, it also presents a credible explanation. In figure 2, I have duplicated the christening data and drawn a horizontal line across the plot through the 1704 data point. In doing so we immediately see that the line goes through exactly one other point—1674. If we went back to Arbuthnot's table we would see that in 1674 the number of christenings of boys and girls were 6,113 and 5,738, exactly the same number as he had for 1704. Thus the 1704 anomaly is likely to be a copying error! In fact, the correct figure for that year is 15,895 (8,153 boys and 7,742 girls), which lies comfortably between the christenings of 1703 and 1705 as expected.

It seems reasonable to assume that Arbuthnot, upon seeing such an unusual data point, would have investigated and, finding a clerical error, would have corrected it. Yet he did not. He did not, despite the fact that when graphed the error stood out, literally, like a sore thumb. Thus we must conclude that he never graphed his data. Why not?

The remarkable answer to this question occupies most of part I of this book. In brief,

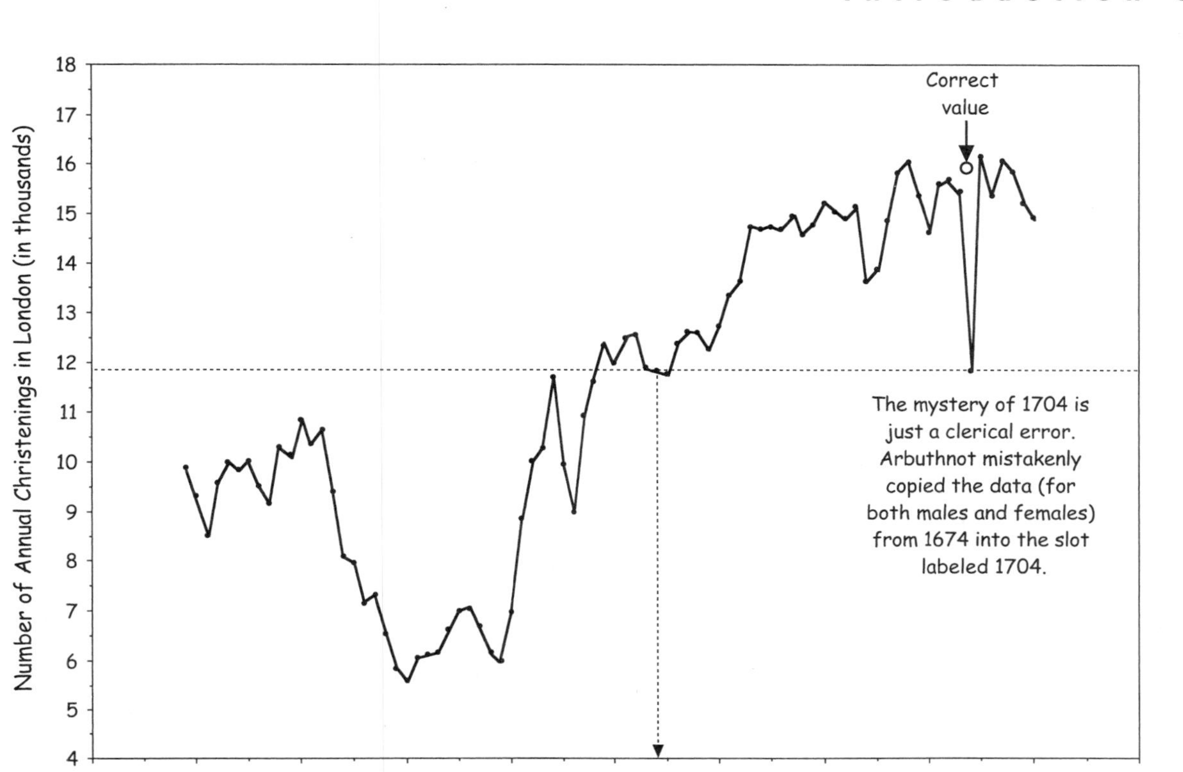

Figure 2. The solution to the mystery of 1704 is suggested by noting that only one other point (1674) had exactly the same values as the 1704 outlier. This coincidence provided the hint that allowed Zabell to track down Arbuthnot's clerical error.

it is that the very idea of graphing data was not yet invented. The story of its invention and inventor is a fascinating one that mixes science and politics, intrigue and scandal, revolution and shopping. The principal actors in this drama include some of the leading scientists of the day as well as a scoundrel of the first order. It also includes a French physician, a scientist of great renown, two signers of the American Declaration of Independence, and America's third president.

Part II illustrates the power of this marvelous invention to help us understand the modern world. It contains examples as disparate as a murder trial in Connecticut, the effect of the Vietnam war on college admissions, faxing policies in Canada, preposterous sports cars, the Boston Marathon, and the perambulations of the stock market over the past century. In each of these situations we see how a graphic depiction of data can help us see things that otherwise would have been as invisible to us as Arbuthnot's clerical error was to him.

Part III of this book looks to the future. I begin with a brief examination of the life and work of John Wilder Tukey, one of the twentieth century's great geniuses. I then describe some of the sorts of graphical methods, closely tied to modern computing, that are likely to become the bedrock of the way that we will be able to understand the torrent of data that has become the standard in our information-laden society.

Superficially, this is a book celebrating data

display. But that phrase has two parts: *data* and *display*. Fine display formats are worthless without important data, and so in celebrating the marvelous invention of data-based graphics we must also save some kudos for those who undertook the often thankless task of gathering the data.* Where would the Johannes Keplers of history have been without the Tycho Brahes?

Graphics evolved because (i) there was a growing recognition that important questions could be answered with data,† (ii) data were being gathered to aid in the quest for such answers, and (iii) graphs were the best way to find both the structure and the surprises hidden in data. The London *Bills of Mortality* were one of the earliest organized efforts to gather extensive longitudinal data in the hope of aiding the government in its search for an effective policy for dealing with public catastrophes such as the bubonic plague. Thus, let me muster a weak thank-you to the plague for providing the immediate motivation for the sort of data gathering that has been the backbone of public health policy in the four hundred years since.

* It is critical, in our efforts to gain a glimpse of the cosmos, that we do not forget the importance of the demanding and sometimes boring intellectual regimen without which discovery is impossible. The famous conductor Pierre Montreux, rich in years and reputation, was once told by a young conductor in one of his master classes at Tanglewood that he, the student, was desperately seeking the "meaning" of the "ineffable essence of Mozart." Montreux congratulated him on his high aim, and then said it would do him no harm, meanwhile, to "learn how to keep the beat."

† In the mid-eighteenth century, Samuel Johnson pointed this out when he said, "To count is modern practice, the ancient method was to guess," but he was far from the first to recognize this, as Seneca too was aware of the difference: "Magnum esse solem philosophus probabit, quantus sit mathematicus." ("The philosopher says the sun is large, but the mathematician measures it." *Epistulae* 88.27)

I William Playfair and the Origins of Graphical Display

The graphic explosion of the nineteenth century that manifested itself in the publication of atlases in all aspects of the observational sciences had its origin in the intellectual turbulence of the eighteenth century. The hundred-year span between 1750 and 1850 saw a shift in the language of science from words to pictures.*

This shift began with the historical time charts of Jacques Barbeau-Dubourg (see chapter 7), which were followed closely by similar efforts from the Scottish philosopher Adam Ferguson and twelve years later from Joseph Priestley (chapter 5). Indeed, this idea proved so useful that Thomas Jefferson even used it to keep track of the price of vegetables in the Washington market (chapter 6). The scientific use of graphic displays had its origin with the Dutch polymath Christiaan Huygens (1629–1693), who developed the first survival chart in 1669, and accelerated when Martin Lister provided graphical summaries of weather data before the Oxford Philosophical Society on March 10, 1683 (chapter 1), which was quickly followed by many others who wanted to understand better the weather data that were now available with the invention of the barometer. All of the plots of these worthy scholars were precursors to the work of the Scot William Playfair (1759–1823), whose *Commercial and Political Atlas of England and Wales* contained no maps but instead beautifully polished versions of most of the common graphical forms in use today. This atlas, in which modern techniques of the graphical presentation of quantitative phenomena emerged fully developed, earned Playfair my nomination for the title of father of modern graphical display. (See chapters 1–4 for more about Playfair's contributions and his remarkable life.) The traditions of graphical display and weather were brought into a wonderful consilience by Francis Galton, who mapped weather reports of December 1861, drawn from all over western Europe, to show for the first time phenomena that had previously only been the hinted at (chapter 8). With Galton, the case for a transition to graphical representation of scientific phenomena was complete.

But others before Galton had already been converted. In 1878, the French physiologist Etienne Marey understood the value of graph-

* After the perfection of engraving in the fifteenth century, scientific documents often included lots of pictures, including abstract diagrams. What did not appear until after Playfair (with a few exceptions such as musical notation and those examples I give in chapter 1) was the diagramming of variates from empirical observations that went beyond the scale translations of space and time as movement in space (Biderman 1990).

ical representation. His graphic schedule of all the trains between Paris and Lyons (see figure I.1) provides a powerful illustration of the breadth of value of this approach. And, on the off chance that someone might have missed the point, he provided an explicit conclusion: "There is no doubt that graphical expression will soon replace all others whenever one has at hand a movement or change of state—in a word, any phenomenon. Born before science, language is often inappropriate to express exact measures or definite relations."[1] Marey was also giving voice to the movement away from the sorts of subjectivity that had characterized prior science in support of the more modern drive toward objectivity.* Although some cried out for the "insights of dialectic," "the power of arguments," "the insinuations of elegance," and the "flowers of language," their protestations were lost on Marey, who dreamed of a wordless science that spoke instead in high-speed photographs and mechanically generated curves, in images that were, as he put it, in the "language of the phenomena themselves."[2]

Historians have pointed out that "Let nature speak for itself" was the watchword of the new brand of scientific objectivity that emerged at the end of the nineteenth century. "At issue was not only accuracy but morality as well: the all-too-human scientists must, as a matter of duty, restrain themselves from imposing their hopes, expectations, generalizations, aesthetics, and even their ordinary language on the image of nature."[3] Mechanically produced graphic images would take over when human discipline failed. Marey and his contemporaries turned to mechanically produced images to eliminate human intervention between nature and representation. "They enlisted polygraphs, photographs, and a host of other devices in a near-fanatical effort to produce atlases—the bibles of the observational sciences"—documenting birds, fossils, human bodies, elementary particles, flowers, and economic and social trends that were certified free of human interference.[4]

The problem for nineteenth-century atlas makers was not a mismatch between world and mind, as it had been for seventeenth-century epistemologists, but rather a struggle with inward temptation. The moral remedies sought were those of self-restraint: images mechanically reproduced and published, warts and all; texts so laconic that they threatened to disappear entirely. Seventeenth-century epistemology aspired to the viewpoint of angels; nineteenth-century objectivity aspired to the self-discipline of saints. The precise observations and measurements of nineteenth-century science required taut concentration endlessly repeated. It was a vision of scientific work that glorified the plodding reliability of the bourgeois rather than the moody brilliance of the genius.†

The graphical representation of scientific phenomena served two purposes. Its primary function was standardizing phenomena in visual form, but it also served the cause of publicity for the scientific community. It preserved what was ephemeral and distributed it to all who would purchase the volume, not just the lucky few who were in the right place at the right time with the right

* Marey's view that graphical display could avoid the problems of language echoed the insight of his ill-fated countryman, Louis XVI, who expressed this view after receiving and reading a copy of Playfair's *Atlas* (see chapter 2).

† Although with such contributors as Condorcet (1743–1794), von Humboldt (1769–1859), and Florence Nightingale (1820–1910), there was certainly room for genius in the eighteenth and nineteenth centuries. Indeed Galton's weather maps, developed in the middle of the nineteenth, show how plodding reliability, when adjoined with moody brilliance, can yield especially fruitful results (chapter 8), yet no one would doubt that Robert Plot was a plodding plotter.

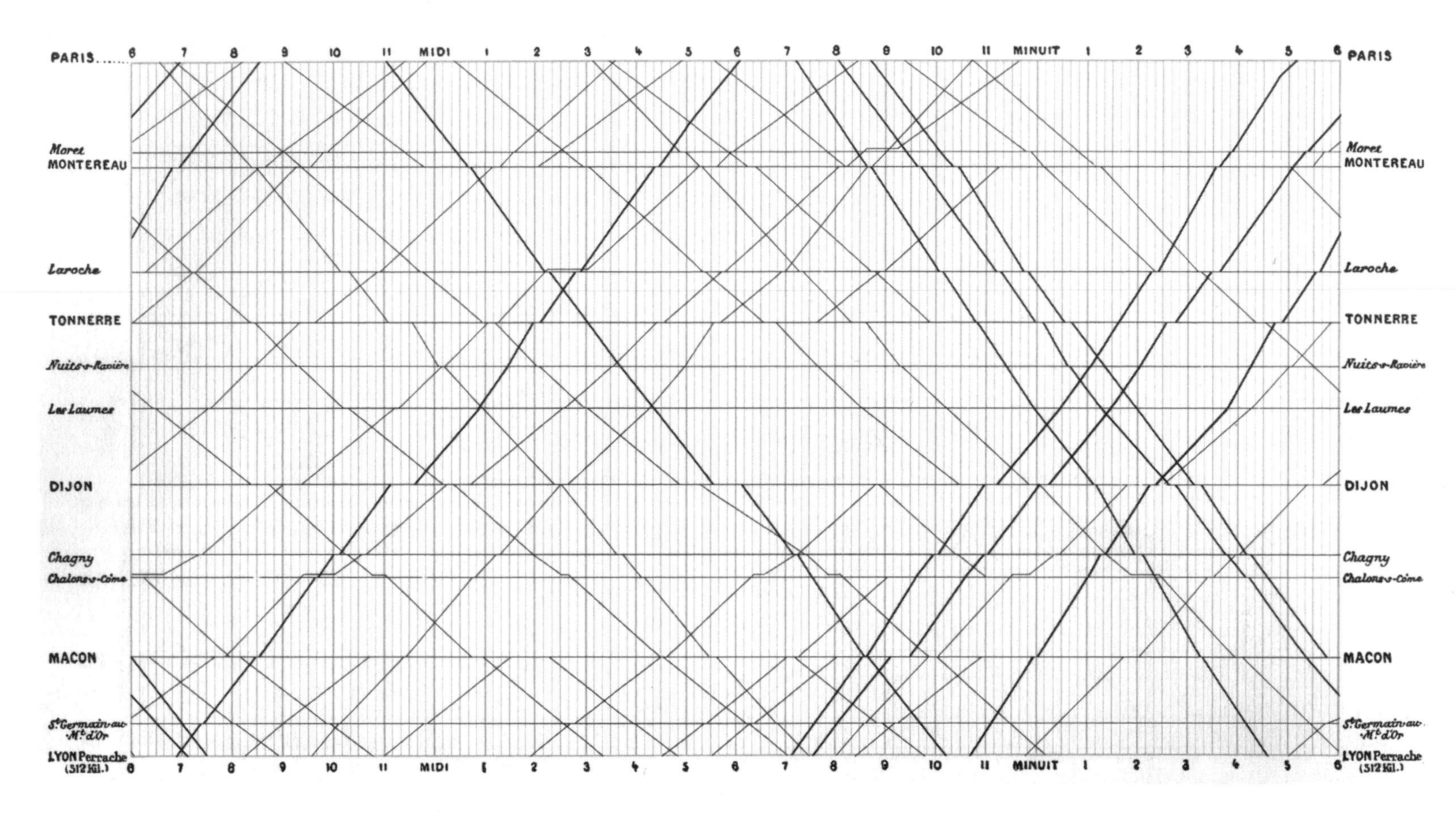

Figure I.1. Etienne Marey's (1878) "graphical train schedule" showing the daily passage of all of the trains between Paris and Lyons. This version, prepared by Edward Tufte (1983), uses a gray grid and is reproduced with his permission.

equipment. And it served the cause of memory, for images are more vivid and indelible than words.

But the graphical display of natural phenomena was viewed as yet more. Marey, in an accompanying note to his design of a portable polygraph, which automatically registered a variety of measures, suggested that through the use of graphics scientists could reform the very essence of scientific research and scientific evidence. "The graphic method translates all these changes in activity of forces into an arresting form that one could call the language of the phenomena themselves, as it is superior to all other modes of expression" (p. iv). Such a language was, for Marey, universal in two senses. Graphical representation could cut across the artificial boundaries of natural languages to reveal nature to all people, and it could cut across disciplinary boundaries to capture phenomena as diverse as the pulse of a heart and the downturn of an economy. Pictures became more than merely helpful tools: they were the words of nature herself.*

* This simple? Perhaps not. An alternative thesis to the one that characterizes science's task as capturing the glorious revelations by nature of her sublime design is one that sees man imposing the order of his senses and his arts upon the unheavenly disorder amidst which he finds himself.

1 Why Playfair?

"Getting information from a table is like extracting sunbeams from a cucumber" (Farquhar and Farquhar).* This evocative indictment of data tables by two nineteenth-century economists comes as no great insight to anyone who has ever tried to draw inferences from such a data display. For most purposes we almost always prefer a graphical representation. Indeed, graphs are ubiquitous now; hence it is hard to imagine a world before they existed. Yet data graphs are a human invention, indeed a relatively modern one. Data-based graphics began to make an appearance in the mid-seventeenth century but their full value and great popularity can be traced to a single event and a single person. The event was the publication, in 1786, of a small atlas describing the imports and exports of England and Wales with their various trading partners. The atlas contained forty-four graphs and no maps. Its author was a twenty-seven-year-old Scot named William Playfair, and his *Commercial and Political Atlas* forever changed the way that we look at data.

William Playfair (1759–1823) worked as a draftsman for James Watt and was the ne'er-do-well younger brother of the well-known scientist John Playfair (1748–1814).† William Playfair is often credited with being the progenitor of modern statistical graphics. Most histories of statistical graphics give him huge credit while acknowledging important graphical work that preceded him.[1] A balanced summary is that he invented many of the currently popular graphical forms,‡ improved the few that already existed, and broadly popularized the idea of graphic depiction of quantitative information. Before Playfair, statistical graphics were narrowly employed and even more narrowly circulated. After him, graphs popped up everywhere, being used to convey information in the social, physical, and natural sciences.

The title comes from Albert Biderman's private characterization of the question that has intrigued him for more than a decade. The intellectual contents of this chapter come principally from two sources: Biderman's 1978 talk at the Leesburg, Virginia, conference on Social Graphics (and its published elaboration, Biderman 1990) and Patricia Costigan-Eaves and Michael Macdonald-Ross's extensive, but as yet unpublished, history of early graphic developments. Some of their material is in Costigan-Eaves and Macdonald-Ross 1990.

* This well-known quotation, though pithy, is somewhat inaccurate. What the brothers Farquhar actually wrote (1891, p. 55) was, "The graphical method has considerable superiority for the exposition of statistical facts over the tabular. A heavy bank of figures is grievously wearisome to the eye, and the popular mind is as incapable of drawing any useful lessons from it as of extracting sunbeams from cucumbers."

† John Playfair's activities were remarkably varied: minister, geologist, mathematician, and professor of natural philosophy at Edinburgh University. In fact, in 1805 William thanked his brother for the idea of using "lines applied to matters of finance" that William used in his 1786 book. We can only speculate why it took him nineteen years to give his brother some credit.

‡ *Invented* is perhaps too strong a term. It may be more accurate to refer to him as an important deployer of graphical forms. He did not invent so much as he permuted and manip-

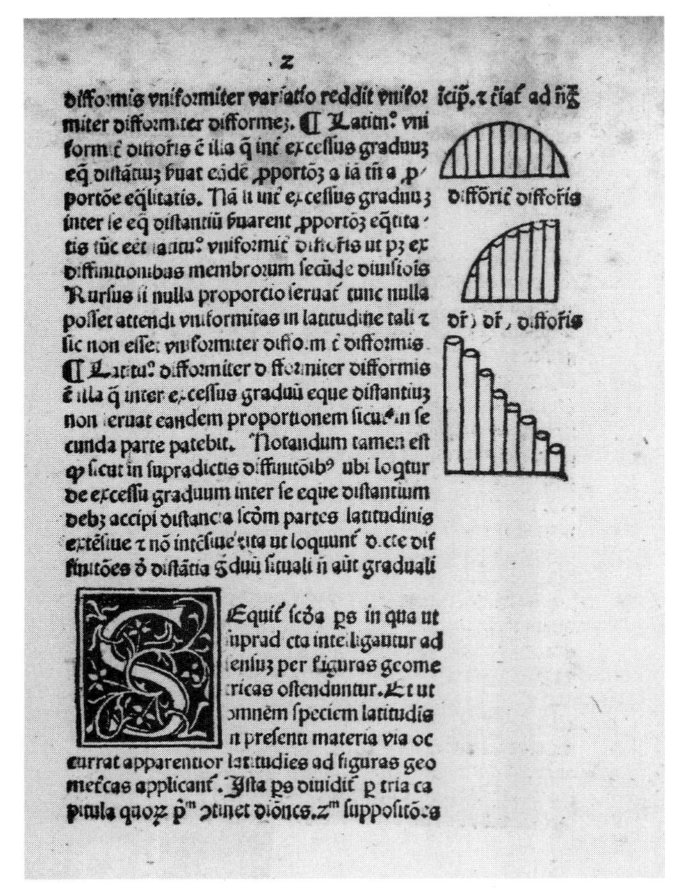
2

difformis vniformiter variatio reddit vnifor- icip. τ ciat ad ...
miter difformiter difforme. ❡ Latitu° vni-
formiť difformis ē illa q̄ inť excessus graduū
eq̄ distātiū hūat eādē pportōe a iā tīi a p-
portōe eq̄litatis. Nā si inť excessus graduū
inter se eq̄ distantiū hūarent pportōe eq̄lita-
tis tūc eēt latitu° vniformiť difformis ut p3 ex
diffinitionibus membrorum secūde diuisiōis
Rursus si nulla proporcio seruať tunc nulla
posset attendi vniformitas in latitudine tali τ
sic non esset vniformiter diffo.m ē difformis.
❡ Latitu° difformiter difformiter difformis
ē illa q̄ inter excessus graduū eque distantiū
non seruat eandem proportionem sicut in se-
cunda parte patebit. Notandum tamen est
q̄ sicut in supradictis diffinitōib9 ubi loq̄tur
de excessu graduum inter se eque distantium
deb3 accipi distancia scd̄m partes latitudinis
extēsiue τ nō intēsiue ita ut loquunť d.cte dif-
finitōes ō distātia s̄duū situali n̄ aūt graduali

Sequiť scd̄a ps in qua ut
suprad cta intelligantur ad
sensū per figuras geome-
tricas ostenduntur. Et ut
omnēm speciem latitudis
in presenti materia via oc-
currat apparentior latitudies ad figuras geo-
metricas applicanť. Ista ps diuidiť p tria ca-
pitula quoꝝ p^m cotinet diōnes. 2^m suppositōes

Figure 1.1. The first page of Oresme's *Tractatus de latitudunibus formarum*, the Padua edition of 1486. This item is reproduced by permission of the Huntington Library, San Marino, California.

Before we meet Playfair, however, it is worthwhile to step back a century or so and examine the attitudes that pervaded scientific investigations. Because natural science originated within natural philosophy, it favored a rational rather than an empirical approach to scientific inquiry. Such an outlook was antithetical to the more empirical modern approach to science, which does not disdain the atheoretical plotting of data points with the goal of investigating suggestive patterns. Graphs that were in existence before Playfair (with some notable exceptions that I will discuss shortly) grew out of the same rationalist tradition that yielded Descartes's coordinate geometry—that is, the plotting of curves on the basis of an a priori mathematical expression. For example, Oresme's "pipes" on the first page of the Padua edition of his 1486 *Tractatus de latitudunibus formarum* (figure 1.1) is often cited as an early example.*

This notion is supported by statements such as that of Luke Howard, a prolific grapher of data in the late eighteenth and early nineteenth centuries who, as late as 1844, apologized for his methodology and referred to it as an "autograph of the curve . . . confessedly adapted rather to the use of the *dilettanti* in natural philosophy than that of regular students."

All the mechanical pieces necessary for data-based graphics were in place long before Playfair. For example, a primitive coordinate system of intersecting horizontal and vertical lines that enable a precise placement of data points was used by surveyors of the Nile flood basin as early as 1400 B.C. A more refined coordinate system was used by Hipparchus (ca. 140 B.C.), whose terms for the coordinate axes translate into Latin as *longitudo* and

---

ulated graphic elements; he varied graphic codes and formats as experiments to try to improve the visual portrayal of quantitative relationships. But there is no doubt that Playfair's use of graphics was more influential than his predecessors'. Part of this must have been because Playfair's graphs were so beautifully produced (compare his line charts with those of Huygens), but more important is the undeniable fact that Playfair published statistical graphics for all to see. Moreover, he did this repeatedly, and with a coherent theme, thus powerfully making the point that the graphical depiction of information can communicate quantitative information in an accurate and relatively painless way.

* Clagett (1968) argued convincingly that this work was not written by Oresme, but probably by Jacobus de Sancto Martino, one of his followers, in about 1390—yet another instance of how surprisingly often eponymous referencing is an indication only of who did not do it (Stigler 1980).

latitudo, to locate points in the heavens. Somewhat later, Roman surveyors used a coordinate grid to lay out their towns on a plane that was defined by two axes. The *decimani* were lines running from east to west, and the *cardi* ran north to south.[2] Many other special-purpose coordinate systems were in wide use before Playfair: for example, musical notation placed on horizontal running lines was in use as early as the ninth century,[3] and the chessboard was invented in seventh-century India.

One of the earliest examples of printed graph paper dates from about 1680.[4] Large sheets of paper engraved with a grid were apparently printed to aid in designing and communicating the shapes of the hulls of ships.* Many historians describe Descartes's 1637 development of a coordinate system as an important intellectual milestone in the path toward statistical graphics.[5] More recent work interprets this in exactly the opposite way—as an intellectual impediment that took a century and a half and Playfair's eclectic mind to overcome.[6]

Although the use of coordinate grids is very old, graphic encoding of information is older still. Paleolithic cave art provides an early and very striking example of graphic display. Some Ice Age bone carvings of animals are intermixed with patterns of dots and strokes that some archeologists have interpreted as a lunar notation system related to the animals' seasonal appearance. These are almost identical in structure, as well as degree of detail, to the engraving on the hull of the Pioneer 10 spacecraft that shows a drawing of a man and a woman along with a simple plotting of the Earth's location by dotted pulsar beams.

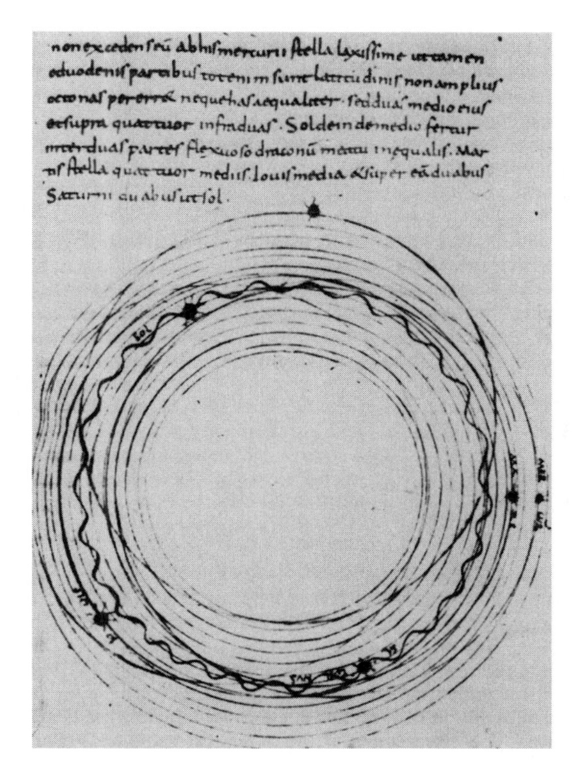

Figure 1.2. Plinian circular diagram of planetary latitudes, early ninth century. Taken from Eastwood (1987), p. 158, figure 6.

So we have the ideas of graphic encoding of information and a coordinate grid system. Why not the plotting of data? Well, some data were plotted. Let us consider three examples.

Example 1. Pliny's Ninth-Century Astronomical Charts

Pliny's (ca. 810) astronomical data were plotted in a roughly circular form (see figure 1.2) corresponding to the varying locations of the bodies in the heavens. But these graphs (astronomical maps) were not as useful as they might have been, because locating a body in the circular path required visually tracking the

* This material is classed in the "collection" category of the British Library with the entry, "A collection of engraved sheets of squared paper, whereon are traced in pencil or ink the curves or sweeps of the hulls of sundry men-of-war."

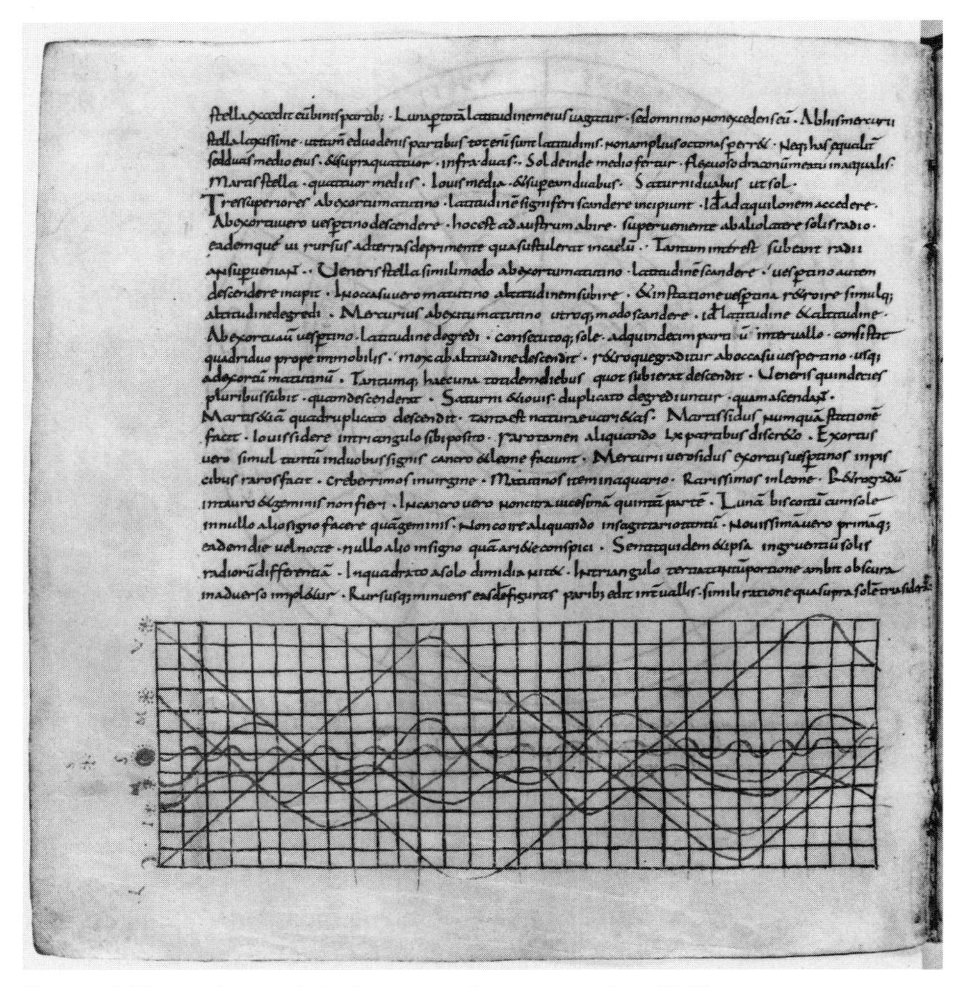

Figure 1.3. Plinian planetary latitudes, portrayed on a rectangular grid. Photo courtesy Burgerbibliothek Bern, ms. 367 f. 24v.

complete cycle. A manuscript originating in Auxerre toward the end of the ninth century contained a scheme that remedied this by transforming the circular grid into a rectangular one (see figure 1.3). The cyclic nature of the orbit is less apparent, but by making explicit the time (horizontal axis) and height above the horizon (vertical axis) it made locating and identifying a heavenly body somewhat easier.*

Example 2. Christiaan Huygens's Seventeenth-Century Survival Charts

On October 30, 1669, the Dutch polymath Christiaan Huygens (1629–1693) received a letter from his brother Lodewijk containing some interpolations of life expectancy data taken from John Graunt's 1662 book *Natural and Political Observations on the Bills of Mortality*. Christiaan responded in letters

* There is actually less here than meets the eye. The horizontal axis is not particularly well defined. It is really an unfolding of the circular version with the horizontal spacing being only roughly related to time

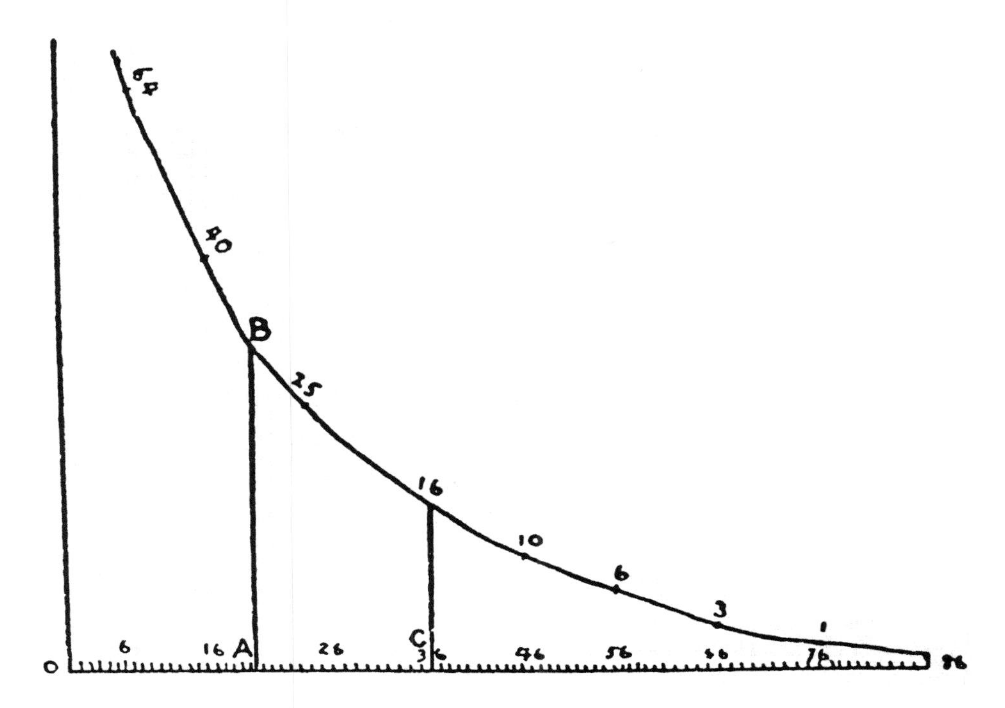

Figure 1.4. Christiaan Huygens's 1669 curve showing how many people out of a hundred survive between the ages of infancy and eighty-six. The data are taken from John Graunt, *Natural and Political Observations on the Bills of Mortality* (1662).

dated November 21 and 28, 1669, with graphs of those interpolations. Figure 1.4 contains one of those graphs showing age on the horizontal axis and number of survivors of the original birth cohort on the vertical axis. The curve drawn was fitted to his brother's interpolations.* The letters on the chart are related to an associated discussion on how to construct a life expectancy chart from this one—that is, analyzing a set of data to gain deeper insights into the subject. Christiaan constructed such a chart and indicated that it was more interesting from a scientific point of view; figure 1.4, he felt, was more helpful in wagering.†

Example 3. Robert Plot's Seventeenth-Century Plots of Barometric Pressure

Good graphs can make difficult problems trivial. We have all become used to weather forecasts that are very accurate and detailed for a

* Huygens's twenty-two-volume *Oeuvres complètes* (1888–1950) contains many other graphical devices to be explored by anyone with fluency in ancient Dutch, Latin, and French. Incidentally, Huygens's graphical work on the pendulum proved to him that a pendulum's oscillations would be isochronic regardless of its amplitude. This discovery led him to build the first clock based on this principle.

† This scooped a 1976 paper by the Chicago statistician Sandy Zabell, whose graphical analysis of the *Bills of Mortality* found inconsistencies, clerical errors, and a remarkable amount of other information, "much of it unappreciated at the time of their publication" (Zabell 1976, p. 27). Zabell's point, though implicit in his paper, is important in this discussion. As we illustrated in one situation in the introduction to this book, this was strong evidence that graphic display was not widely available. For had they been seen, these errors, which could not be missed with any sort of competent display, would have been discovered and eliminated.

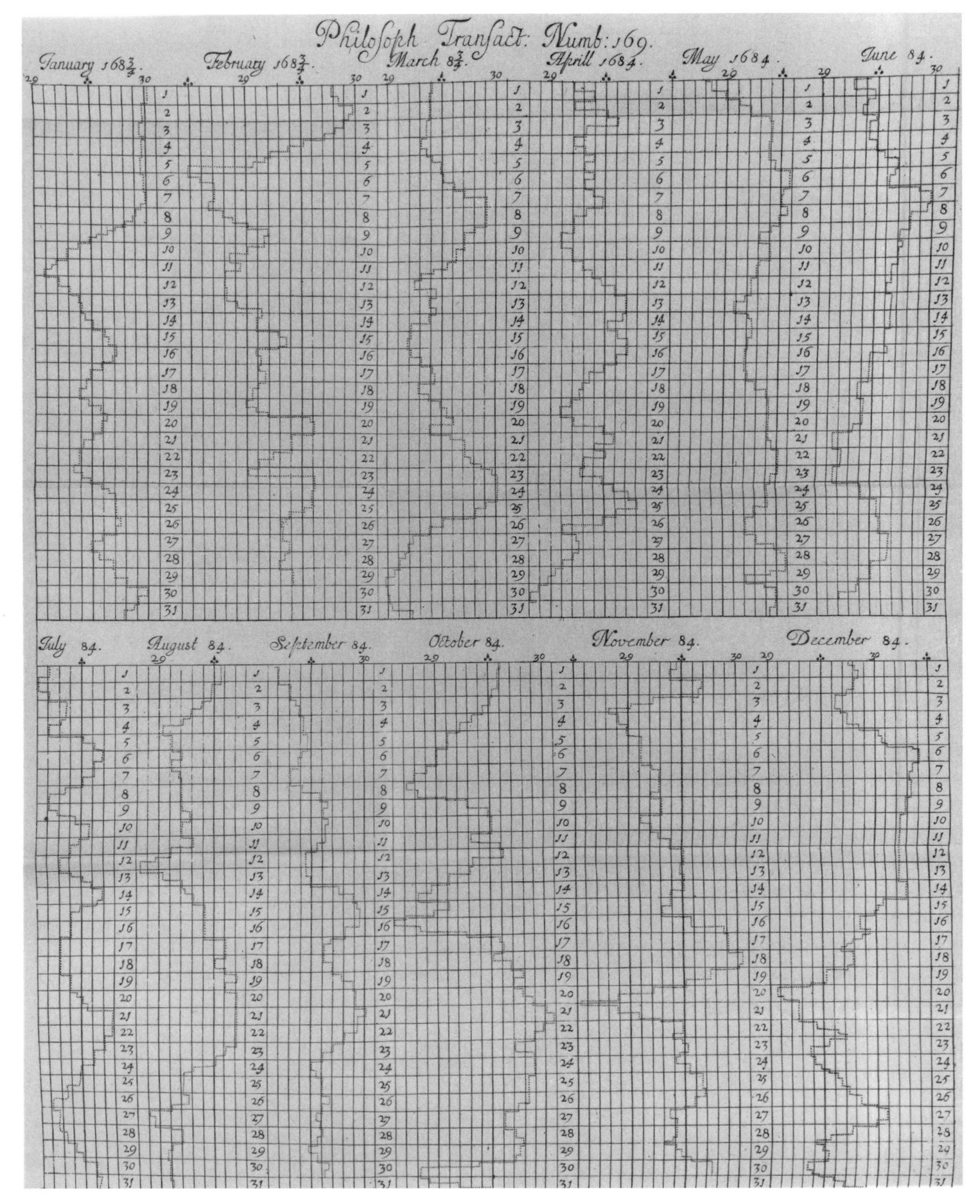

Figure 1.5. Robert Plot's (1685) "History of the Weather" recording of the daily barometric pressure in Oxford for the year 1684. Appears in *Philosophical Transactions* and is based on the original work of Martin Lister. Photo © The Royal Society.

day or two and pretty good for as far in advance as a week. I used to think that this was due to the increasing sophistication of complex prediction models.* But then I noticed the weather maps shown on every news broadcast. Using a model of no greater sophistication than that employed by Benjamin Franklin (weather generally moves from west to east), I was able to predict that the area of precipitation currently over Ohio would be hitting New Jersey by tomorrow and would stay over us until the weekend. Any fool could see it. The improvement in forecasting has not been entirely due to improvements in the mathematical models of the weather. The enormous wealth of radar and satellite data summarized into a multicolored and dynamic graphic can turn anyone into an expert.

The path to modern weather graphs is more than three hundred years long. The barometer was developed in 1665. Robert Plot recorded the barometric pressure in Oxford every day in 1684 and summarized his findings in a remarkably contemporary graph (figure 1.5) that he called a "History of the Weather."

He sent a copy of this graph with a letter to Martin Lister† in 1685 with a prophetic insight:

> For when once we have procured fit persons enough to make the same Observations in many foreign and remote parts, how the winds stood in each, at the same time, we shall then be enabled with some grounds to examine, not only the coastings, breadth, and bounds of the winds themselves, but of the weather they bring with them; and probably in time thereby learn, to be forewarned certainly, of divers emergencies (such as heats, colds, dearths, plague, and other epidemical distempers) which are not unaccountable to us; and by their causes be instructed for prevention, or remedies: thence too in time we may hope to be informed how far the positions of the planets in relation to one another, and to the fixed stars, are concerned in the alterations of the weather, and in bringing and preventing diseases and other calamities . . . we shall certainly obtain more real and useful knowledge in matters in a few years, than we have yet arrived to, in many centuries.

With so many predecessors, why have I chosen Playfair as my candidate for the father of modern graphical display? Although the arguments will build in subsequent chapters, a visual comparison of a sampling of Playfair's plots with any of those that came before him makes clear the qualitative jump that Playfair's work represents. Ian Spence points out that Playfair was the first to use hachure, color, and area to represent quantities in a systematic way. Moreover, Playfair published these forms in widely circulated volumes. Most of all, however, Playfair's graphs provided proof that the presentation of evidence could be beautiful.

As but one such example, compare Playfair's harmonious depiction of more than a century of data on England's national debt

* It is true that models are more sophisticated than they were in the past. I was enormously impressed when some surprising turns in a hurricane's path were predicted well in advance, but such models seem to be needed no more often than seldom.

† The origin of the graphical depiction of weather data, sadly, for the obvious eponymous glory, rests not with Plot but rather with Lister, who presented various versions of graphical summaries of weather data before the Oxford Philosophical Society on March 10, 1683, and later in the same year presented a modified version to the Royal Society. Plot was not the only one enthusiastic about Lister's graphical methods. William Molyneux was so taken that he had an engraving made of the grid, and he faithfully sent a "Weather Diary" monthly to William Musgrave. One of Molyneux's charts was reproduced in Gunther (1968).

(figure 1.6) with any of the graphs produced previously by others. The viewer's eye is drawn from the soaring debt to the vertical lines that communicate the events that presaged a change in the debt. The viewer's mind cannot avoid making the causal inference suggested. Nothing that had been produced before was even close. Even today, after more than two centuries of graphical experience, Playfair's graphs remain exemplary standards for clear communication of quantitative phenomena.

It now seems wise to recapitulate the argument: Graphical forms were available before Playfair, but they were rarely used to plot empirical information. I argue that this was because there was an antipathy toward the empirical approach. This suggestion is supported by statements such as that made by Luke Howard. But at least sometimes when data were available (for example, Pliny's astronomical data, Graunt's survival data, Plot's weather data, and several other admirable uses), they were plotted. Could it be that the exponential increase in the use of graphics after the publication of Playfair's *Atlas* was merely concomitant to the exponential growth in the availability of data? Or did the availability of graphic devices for analyzing data encourage data gathering? And why Playfair? Was he merely at the cusp of an explosion in data gathering and so his graphic efforts appear causal? Or did he play an important role in that explosion?

The consensus of scholars is that until Playfair "many of the graphic devices used were the result of a formal and highly deductive science. . . . This world view was more comfortable with an arm-chair, rationalistic approach to problem-solving which usually culminated in elegant mathematical principles" often associated with elegant geometrical diagrams.[7] The empirical approach to problem solving, a critical driving force for data collection, was slow to emerge. But the empirical approach began to demonstrate remarkable success in solving problems, and with improved communications,* the news of these successes and hence the popularity of the associated graphic tools began to spread quickly.

So the picture is almost complete. The fundamental tools for the graphical display of data were available; there was an increase in the acceptance of an empirical approach to science as an important part of the scientific process; data were being gathered in greater and greater quantities; and the success of empirical procedures in solving important practical problems was being more widely communicated. This explains the growth of the graphical method, but still leaves the initial question, "Why Playfair?"

We are accustomed to intellectual diffusion taking place from the natural and physical sciences into the social sciences; certainly that is the direction taken for both calculus and the scientific method. But statistical graphics in particular, and statistics in general, went the reverse route. Although, as we have seen, there were applications of data-based graphics in the natural sciences, only after Playfair applied them in the social sciences did their popularity begin to accelerate. Playfair should be credited with producing the first chartbook of social statistics; indeed, publishing an *Atlas* that contained not a single map is one indication of his belief in the methodology (to say nothing of his chutzpah). Playfair's work was immediately admired, but emulation, at least

* The first encyclopedia in English appeared in 1704. The number of scientific periodicals began a rapid expansion at the end of the eighteenth century; between 1780 and 1789 twenty new journals appeared, and between 1790 and 1800 twenty-five more (McKie 1972)

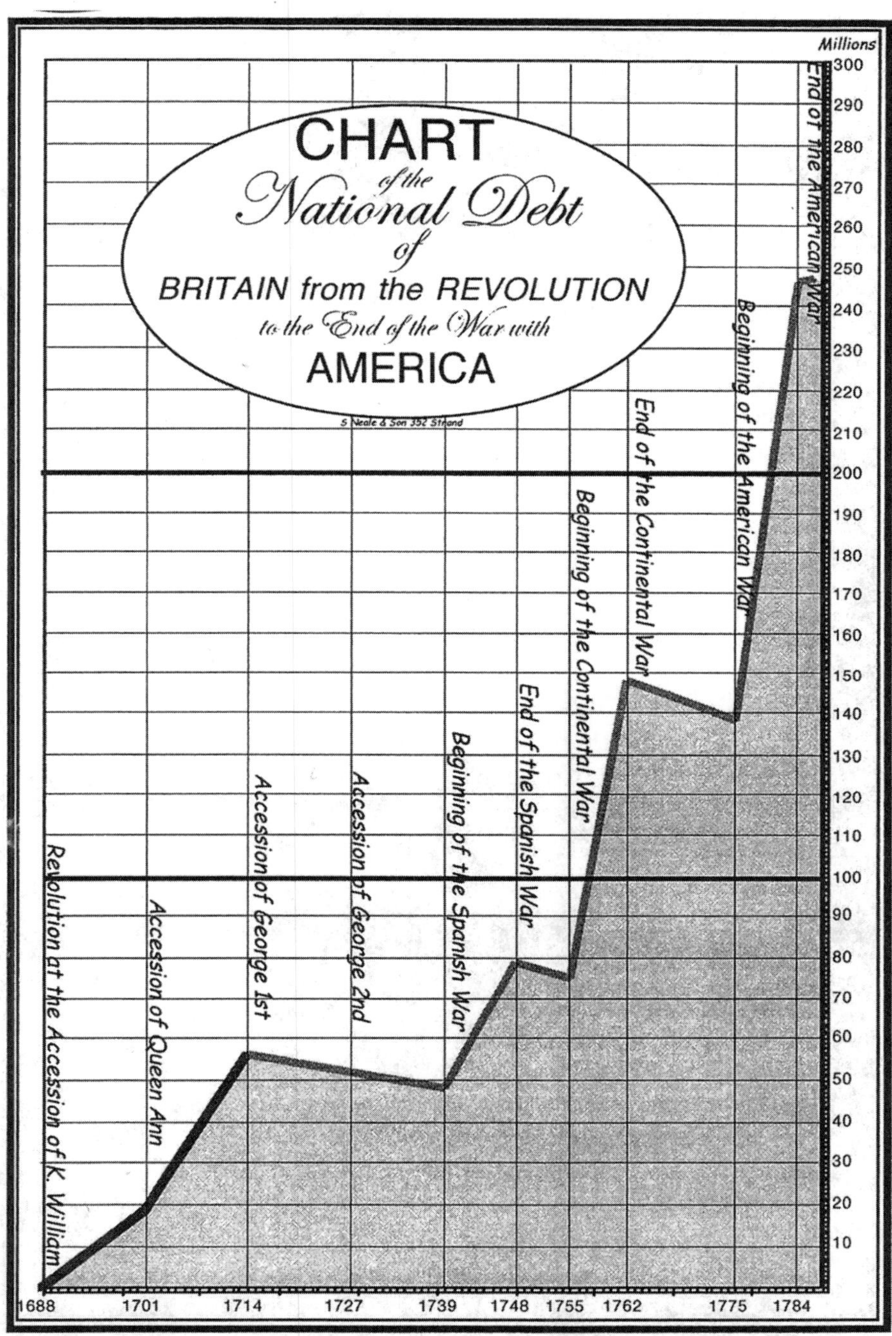

Figure 1.6. A close facsimile of William Playfair's plot of England's national debt from 1688 until 1786. It appeared in his *Commercial and Political Atlas* and accompanied his discussion arguing against the British government's policy of financing its colonial wars through debt.

in Britain, took a little longer (graphics use started on the continent a bit sooner). Interestingly, one of Playfair's earliest emulators was the banker S. Tertius Galton (the father of Francis Galton, and hence the biological grandfather of modern statistics), who in 1813 published a multiline time-series chart of the money in circulation, rates of foreign exchange, and prices of bullion and of wheat.* The relatively slower diffusion of the graphical method back into the natural sciences provides additional support for the hypothesized bias against empiricism there. The newer social sciences, having no such tradition and faced with both problems to solve and relevant data, were quicker to see the potential of Playfair's methods.

Playfair's graphical inventions and adaptations look contemporary. He invented the statistical bar chart out of desperation, because he lacked the time-series data required to draw a line showing the trade with Scotland and so used bars to symbolize the cross-sectional character of the data he did have. Playfair acknowledged Priestley's priority in this form, although Priestley used bars to symbolize the life spans of historical figures in a time line.[8] (See chapters 5, 6, and 7 for more on the fascinating history of time lines and graphical display of historical data.)†

Playfair's role was crucial for several reasons, but it was not for his development of the graphical recording of data; others preceded him in that. Indeed, in 1805 he pointed out that as a child his brother John had him keep a graphical record of temperature readings. But Playfair was in a remarkable position. Because of his close relationship with his brother and his connections with James Watt, he was on the periphery of science. He was close enough to know of the value of the graphical method, but sufficiently detached in his own interests to apply them in a very different arena—that of economics and finance.

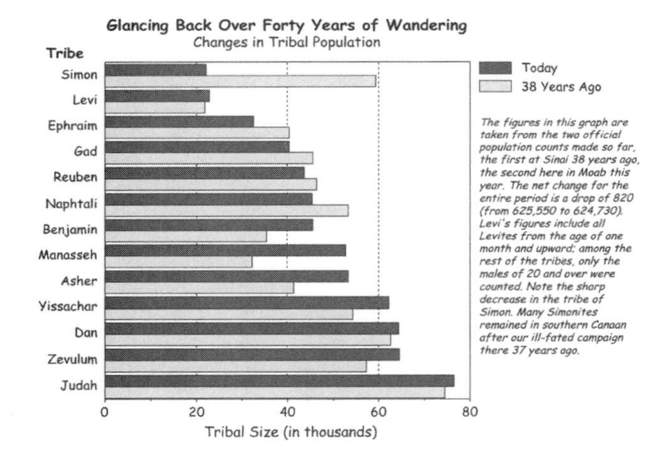

Figure 1.7. A translated and computer-enhanced reproduction of perhaps the earliest statistical graphic yet uncovered. It was apparently constructed about 1400 B.C. and was preserved in a sealed ceramic container in the Qumran caves. It was purchased by the author from Moishe the mapman at his Dead Sea antiquities stall in 1991.

* Biderman (1978) pointed out that, ironically, Galton's chart predicted the financial crisis of 1831 that created a ruinous run on his own bank.

† Priestley's use of the bar as a metaphor is somewhat different than Playfair's in that the data were not really statistical. A much earlier precedent has been recovered from its resting place in the Qumran caves abutting the Dead Sea. The graphic dates from approximately 1400 B.C. and was prepared as a summary of population changes in the twelve tribes of Israel as they emerged from their almost four decades of wandering in the Sinai after their exodus from Egypt, which began in April 1446 B.C. A faithful copy of this bar chart, with the captions and legends translated from Aramaic, is reproduced here as figure 1.7. Some aspects of this historic figure have been computer-enhanced for better reproduction. Note that it presages Huygens in subject matter and Playfair in form.

These areas, then as now, tend to attract a larger audience than matters of science, and Playfair was adept at self-promotion.

In a review of his 1786 *Atlas* that appeared in the *Political Herald*, Dr. Gilbert Stuart wrote,

> The new method in which accounts are stated in this work, has attracted very general notice. The propriety and expediency of all men, who have any interest in the nation, being acquainted with the general outlines, and the great facts relating to our commerce are unquestionable; and this is the most commodious, as well as accurate mode of effecting this object, that has hitherto been thought of. . . . To each of his charts the author has added observations [that] . . . in general are just and shrewd; and sometimes profound. . . . Very considerable applause is certainly due to this invention; as a new, distinct, and easy mode of conveying information to statesmen and merchants.

Such wholehearted approval rarely greets any scientific development. Playfair's adaptation of graphic methods to matters of general interest provided an enormous boost to the popularity of statistical graphics. His energy and artistic sense showed themselves in the forty color charts in his initial *Atlas*. The size of the undertaking required to produce such a book indicates Playfair's deep understanding of the power of the graphical method. His energy and skill as a draftsman, coupled with that understanding, led him to communicate his enthusiasm both widely and effectively. However, to be able to focus on graphics when the prevailing view of science looked upon such an approach as generally illegitimate requires a willingness to go against the tide—indeed, perhaps even taking joy in being an iconoclast. The events described in the next two chapters illuminate this aspect of Playfair's personality.

In *Kagemusha*, a film by the great Japanese director Akira Kurosawa, a legendary warlord is mortally wounded. The warlord's staff finds a petty thief, who bears a remarkable physical resemblance to the fallen leader, to substitute for him. With the substitute in place, the political strategy evolved by the dead warlord succeeds in his absence. In this examination of the question "the man or the moment?" Kurosawa clearly favors the latter. The Playfair enigma represents another instance of this great theme, although unlike Kurosawa's fictional situation, the more limited information available to us does not allow unambiguous conclusions.

2 Who Was Playfair?

Ian Spence and Howard Wainer

> To count is modern practice, the ancient method was to guess.
> —Samuel Johnson, *A Journey to the Western Islands of Scotland*

The Cartesian tradition of graphical representation of mathematical functions worked against the use of graphs to depict empirical regularities. The switch to the view that a graph can help us formulate an understanding of nature by plotting data points and looking for patterns required, in Thomas Kuhn's terms, a change in paradigm. A person who might effect such a change would not only have to be in the right place at the right time, but would also have to be an iconoclast of the first order. In all aspects William Playfair fits the bill. In this chapter we sketch some of the events of his life with the aim of introducing you to the sort of man William Playfair was and thus helping to illustrate why he (and not someone else) invented modern graphical display. The chapter concludes with a short memoir of Playfair's that, as well as any, characterizes his personality and sense of values.

William Playfair was born in 1759 in Scotland during the Enlightenment—a golden age in the arts, sciences, industry, and commerce. He died in London in 1823, after a life that was eventful but unmarked at the time by any apparently significant or memorable contribution. He made little impression in his native land, and his impact was only slightly greater in England and France. Yet he is responsible for inventions familiar and useful to us all: he was the first to devise and publish all of the common statistical graphs—the pie chart, the bar chart, and the statistical line graph. He invented a universal language useful to science and commerce alike and, although most of his contemporaries failed to grasp its significance, he had no doubt that he had forever changed the way we would look at data. However, it took almost a century before his invention was fully accepted.

Despite the importance of Playfair's innovations, his name is largely unknown, even to professional statisticians, and those who have heard of him know little of his life. One might expect a life of the inventor of statistical graphs to make dull reading, but Playfair pursued a variety of careers with such passion, ambition, industry, and optimism that even without his great inventions, he would be judged a colorful figure. He was, by turns, millwright, engineer, draftsman, accountant, inventor, silversmith, merchant, investment broker, economist, statistician, pamphleteer, translator, publicist, land speculator, convict,

This chapter draws heavily on a paper by Spence and Wainer (1997a).

banker, ardent royalist, editor, blackmailer,* and journalist. Some of his business activities were questionable, if not downright illegal, and it may fairly be said that he was something of a rogue and scoundrel.

William Playfair was the fourth son of the Reverend James Playfair of the parish of Liff and Benvie near the city of Dundee, Scotland. His father died in 1772, leaving the eldest brother, John, to care for the family. John was subsequently to become one of Britain's foremost mathematicians and scientists as the professor of natural philosophy, mathematics, and geology at Edinburgh University. After an apprenticeship with Andrew Meikle, the inventor of the threshing machine, William became a draftsman and the personal assistant to the great James Watt at the steam engine manufactory of Boulton and Watt at Birmingham in 1777. Thus, William's scientific and engineering training was at the hands of the leading figures of the Enlightenment and the Industrial Revolution. On leaving Boulton and Watt in 1782, Playfair set up a silversmithing business and shop in London, but the venture failed. Seeking his fortune and hoping to apply his engineering skills to better effect in a developing French economy, he moved to Paris in 1787. He was involved in more than just business in Paris, which was about to undergo revolutionary change. Playfair has been named as one of the approximately twelve hundred inhabitants of the St. Antoine quarter who formed themselves into a militia and assisted in the storming and capture of the Bastille.†

In February 1791, he rescued the well-known ex-judge Duval d'Esprémesnil from the mob in the Palais Royal Gardens. As well as being a friend, Duval was a subscriber to the Compagnie du Scioto, in which Playfair was a principal. This doomed American-French scheme was devised to settle European migrants at the confluence of the Ohio and Scioto rivers. As the French Revolution became more violent, lurching toward the Terror, Playfair became increasingly and vocally disenchanted with the revolutionaries and was forced to quit Paris, narrowly escaping the wrath of Bertrand Barère‡ and who knows what fate.

From the mid-1790s on, Playfair made his living principally as a writer and pamphleteer, although he also applied his engineering skills working as a gun carriage maker and developing the occasional new mechanical invention. Disenchanted by his experiences in Paris, he argued vehemently against the excesses of the French Revolution and wrote frequently on the topic of British policy toward France. He claimed credit for warning of Napoleon's escape from Elba—a warning that the British government unfortunately chose to ignore. His illustrated *British Family Antiquity* was a massive nine-volume undertaking in which he catalogued the peerage and baronetage of the United Kingdom—a work principally designed to raise money by subscription. He

* See chapter 3 for more details of this fascinating, albeit unpleasant, episode in Playfair's life.

† Playfair's participation is described in Georges Lecocq's *Prise de la Bastille* (1881), which was cited by the *Dictionary of National Biography* in its discussion of Playfair, but is not verified by any other authority. This suggests that we should hesitate in placing Playfair at this scene. For example, Thomas Carlyle does not mention Playfair in his extensive description of the taking and razing of the Bastille; since Carlyle took classes from John Playfair and very likely knew William also, this seems especially significant.

‡ Bertrand Barère (1755–1841) was a revolutionary and regicide who, although originally a moderate in the National Convention, helped form the Committee of Public Safety. He was imprisoned after the fall of Robespierre (1794) but escaped. He later served under Napoleon and was exiled with him, only to return to Paris under an amnesty in 1830.

dabbled in journalism, editing more than one periodical, the best known of which may be the *Tomahawk*. After the restoration of the Bourbons, he returned to Paris as editor of the periodical *Galignani's Messenger*, but in 1818 his comments on a duel between Colonel Duffay and the Comte de St. Morys were felt to be libelous by the widow and daughter of the latter and led to prosecution. Playfair was sentenced to three months imprisonment and to pay a fine and damages. To avoid incarceration, he fled France and spent his few remaining years in London writing pamphlets and doing translations.

The last two years of his life saw a renewed interest in economics, and his final publications contain several charts, including one or two rather fine examples that combined the line graph, bar chart, and chronological diagram in a single chart. His interest in agricultural matters was the stimulus for his last two works, which examined the difficulties experienced by English farmers in the early nineteenth century.

A theme throughout Playfair's life was his practical inventiveness. He took out several patents, mostly involving machines for metalworking, but he also proposed innovations such as modifications to the bows of ships to make them faster and he is on record as the inventor of the first mass-produced silver-plated spoon. When he was confronted with a problem, he would frequently offer a practical solution, often ingeniously adapting or exploiting the work of others. Of equal importance was his insistence on recording his inventiveness. For example, about his arrival in Germany in 1793 he wrote,

> When I was in Germany I was surprised that in a country where the milk is excellent the butter was little better than common grease without anything either of the colour or taste that good butter possesses. But one day in changing horses where the post master spoke a little French and had a farm I asked to see the dairy when I found that the milk was kept in deep narrow jars about three feet deep and eight or nine inches wide. The cream that rose to the top was about three inches in depth before it was taken off and though not quite rancid had a disagreeable smell. I advised him to get wide shallow vessels and keep them very clean but he smiled as if I knew nothing of the business. I asked him if the Dutch butter was not better than theirs. He owned it was. I apprised him that the Dutch milk was not so good as the German and that the excellence of the Dutch butter proceeded from the better mode of keeping the milk. He did not attempt to answer my reasoning but gave his head a significant shake and no doubt unless the French soldiers carried them into a better method in Germany they still persist in the same.[1]

In his political and economic writings, he often used numerical examples and calculations to make a point. He found that making sense of empirical information was aided enormously through the use of statistical graphics. He used, refined, and adapted those graphical forms that were known to him and invented others. His contributions to the development and demonstrations of the use of statistical graphics remain his life's principal accomplishment.

In 1786, he published his *Atlas*, which contained forty-four graphs (and no maps). This was the first description of his graphical inventions and is the first major work of any kind to contain statistical graphs. It met with limited initial success in England, but fared quite differently in France. Playfair reports,

> When I went to France, 1787, I found several copies there, and, amongst others, one which had been sent by an English nobleman to the Monsieur de Vergennes, which copy he

> presented to the king, who, being well acquainted with the study of geography, understood it readily, and expressed great satisfaction. This circumstance was of service to me, when I afterwards solicited an exclusive privilege for a certain manufactory, which I obtained. The work was translated into French, and the Academy of Sciences, (to which I was introduced by Mons. Vandermond,) testified its approbation of this application of geometry to accounts, and gave me a general invitation to attend its meetings in the Louvre; and at the same time did me the honour of seating me by the president during that sitting.*

In his unpublished memoirs, Playfair adds, "As his majesty made Geography a study, he at once understood the charts and was highly pleased. He said they spoke all languages and were very clear and easily understood."

* This note appears on page iv of Playfair's pamphlet *For the Use of the Enemies of England: A Real Statement of the Finances and Resources of Great Britain* (1796); he repeated it verbatim on page 6 in the introduction to *Lineal Arithmetic* (1798), and again on page ix in the introduction to the third edition of his *Commercial and Political Atlas* (1801). He was obviously very fond of the royal treatment he received in France

3 William Playfair: A Daring Worthless Fellow

Ian Spence and Howard Wainer

In 1816, William Playfair sought to blackmail Lord Douglas. The latter had been at the focus of the longest legal proceeding in Scottish history, known familiarly as the Douglas Cause. The events precipitating the Cause occurred around the time of Douglas's birth, several years before Playfair's own, and are at the heart of the attempted extortion. The episode illuminates an aspect of Playfair's personality that arguably played a fundamental role in both the invention and the delay in adoption of statistical graphs.

Between 1786 and 1801, Playfair invented the pie chart and was also first to apply the line and bar graph to economic data. Hence he invented or perfected three of the four fundamental types of statistical graphs; the lone exception, the scatterplot, did not appear until the middle of the nineteenth century.[1] Although an enthusiastic advocate of his graphical inventions, Playfair found acceptance lacking, and it was not until almost a century later that the value of his work was realized. By the early twentieth century, his graphs had been accepted so completely that the name and contribution of the original inventor was largely forgotten. Today, proponents of statistical graphs generally think them to be so simple and obvious that almost anyone could have invented them, given sufficient need and interest. Indeed, their apparent simplicity accounts for much of their appeal. Thus we give scarcely a thought to these inventions or their inventor. Familiarity has blunted our sense of the significance of graphs and in consequence has diminished the importance of their creator.

Nowadays, with illustration at the core of almost all forms of communication, it is difficult to comprehend that an eighteenth-century academic would have shown no enthusiasm for the use of pictures to communicate economic data. Readers were accustomed to persuasion by rhetorical means, and illustration in serious writing was viewed with suspicion. It would have been unthinkable to introduce pictorial material to bolster an argument where tabular presentation would have been seen as sufficient and certainly more accurate. Statistical graphs could have been invented and published in the late eighteenth century only by an individual who possessed a complex of technical skills as well as a disdain for convention—the latter was as necessary as the former. The promulgation of displays for statistical data was expedited by the improvidence and brashness of their inventor, even though such unattractive traits

This chapter is based on an article by Spence and Wainer (1997b).

probably contributed materially to the widespread and long-lasting indifference to Playfair's innovations.

It may seem curious that the inventor of statistical graphs should have been involved in an activity so sordid as attempted blackmail, but such behavior may not have been uncharacteristic. William Playfair is known to have been involved in many questionable enterprises contrived to enrich, but almost invariably without success. During his early employment it was alleged that he attempted to register patents claimed as his own invention, when the ideas belonged to James Keir or to others at Boulton and Watt. Neither James Watt nor Matthew Boulton supported him against Keir, even though Playfair wrote self-serving letters of justification to both his employers. A few years later in Paris he was suspected of embezzlement from the Franco-American Scioto land company, thus accelerating the company's collapse, but it is likely that simple bungling by Playfair and Joel Barlow, the American representative who led the French company, caused the failure. In 1793, while in Frankfurt, Playfair heard a description of the semaphore telegraph from a French émigré. The following day he built a model of the apparatus and sent it to the Duke of York. Subsequently, he claimed to have invented semaphore signaling when he had merely introduced a copy of a German device to England. In London in 1791, there were irregularities in a failed partnership with John Caspar Hartsinck in the Cornhill Security Bank (whose business practices were modeled on schemes that Playfair had seen introduced in Paris), including an entanglement with the Bank of England that almost led to prosecution. However, he was subsequently convicted at the King's Bench in 1805 of an unrelated fraud. In later life he was found guilty of libel in Paris, causing him to flee France.

These few glimpses of the freewheeling entrepreneur and confidence man suggest, but do not positively confirm, that Playfair's ethical standards were altogether different from those of his highly respected brother John, a clergyman and professor of mathematics and geology. We lack sufficient particulars to be sure that Playfair's intentions were less than honorable. Simple naïveté, miscalculation, or poor business sense could have been the cause of his misfortunes. We do not lack documentation, however, in the matter of the attempted extortion of Lord Douglas, the details of which have come down to us in Playfair's own words in unpublished and relatively inaccessible correspondence still held by the family of the intended victim.

William Playfair tried to broker the sale of some papers—potentially damaging to Lord Douglas—alleged to provide new information relating to the disputed inheritance of the Douglas estates upon the death of the childless Duke of Douglas. The rival Duke of Hamilton contended that the then-young Archibald Douglas, nephew of the old Duke of Douglas, was no blood relative but had been purchased as an infant in Paris. Notwithstanding, the lengthy legal proceedings were finally resolved in favor of Archibald Douglas. As ultimately becomes clear in the blackmailing correspondence of a half-century later, Playfair had no new damaging information but had devised the scheme to extort money. These letters are perhaps the only direct evidence that Playfair was prepared to follow an unscrupulous and felonious course for material gain. We do not, of course, claim that a criminal predisposition was necessary for the invention of statistical graphs, but we do contend that brashness and an unconcern for the reproof of others would have been highly advantageous. The correspondence gives convincing evidence that such traits were part of Playfair's makeup.

The Playfair-Douglas Correspondence

It would be hard to invent a story more strange or romantic than that of the great Douglas Cause, which occupied Scottish and French courts for several years in the mid-1700s, before final settlement in the House of Lords. The Cause involved the two most powerful families in Scotland, the Douglases and the Hamiltons, a huge inheritance, and a mystery concerning alleged imposture that has not been satisfactorily resolved to this day. The tale concerns two powerful dukes—the one Archibald Douglas, aging, without heir, and said to be half-mad, and the other the Duke of Hamilton—in line to inherit the Douglas titles and estates. But the leading players in the drama were the Duke of Douglas's sister, Lady Jane, and her husband, Colonel John Stewart, partners in a late-life marriage contracted in secret. The groom, a swashbuckling, spendthrift, sixtyish Jacobite adventurer of dubious reputation, was as different from the Presbyterian Lady Jane, and as unsuitable a candidate for marriage, as could be imagined. Small wonder that the existing rift between brother and sister deepened.

Once married, the couple left for an extended period on the Continent. In 1746, the wounds opened by the second Jacobite rebellion were still raw, and the newlyweds would have expected to find Aix-la-Chapelle a more sympathetic milieu than Scotland. After almost two years and several moves, Lady Jane, then fifty, gave birth to male twins in Paris under mysterious circumstances, and named the elder twin Archibald, like her brother. After her return a few years later, and after much effort, Lady Jane finally succeeded in persuading her brother to name the elder twin as heir, thus retaining the succession in the Douglas family. Despite much gossip regarding the peculiar circumstances of the accouchement, it was not until Douglas's death that the issue was forced into the open—the Hamilton faction contended that the children had been purchased in Paris by John Stewart, and therefore the estates should fall to the Duke of Hamilton. Thus began a seven-year legal battle that not only uncovered, but also provoked, all manner of deceit, fabrication, forgery, and bribery. Although initially successful in the Scottish Court of Session, the Hamiltons lost on appeal to the House of Lords, where the judgment favored Lady Jane Douglas as the natural mother of young Archie, who inherited the Douglas estates (but not the dukedom).* This summary is insufficient to do justice to this extraordinary affair, and the reader is referred elsewhere for extended accounts.[2]

Playfair enters the tale a half-century later, in 1816, when he began a correspondence with Archibald Douglas in which he tells him of the existence of papers that he, Playfair, had uncovered in Paris. He implies that if these papers should fall into the hands of others less scrupulous than himself, the case could be reopened. Playfair offers to obtain the papers for Douglas for about fifty thousand dollars (in today's terms). After extended correspon-

* After these tremendous exertions to protect the Douglas line, fate dealt a curious hand. Although young Archie had twelve children, of the eight males none had children and of the four girls only one had a child, a female. Through the marriage of this girl, the estates passed to the duke of Buccleuch, and in the next generation to the Home family, there being no male heir from the Buccleuch union, and there the line rests with the Douglas-Home family.

dence, Lord Douglas, on the advice of his solicitor, declined to yield to the attempted extortion.

Character and the Invention and Promulgation of Graphs

Thanks to his mathematician-scientist brother John—and to his masters, the Scottish engineers Andrew Meikle and James Watt—Playfair had the intellectual and technical skills to invent, produce, and publish statistical graphs. His exposure to commerce in the world's first manufactory under the entrepreneurial Matthew Boulton, James Watt's partner, inspired him to write on economic issues. All these were essential to the invention but they were not enough. There is ample evidence of strong opposition to illustration in serious writing among eighteenth-century academics; although they may not have been averse to the use of charts and diagrams in their laboratories, they were not keen to see these "trifles" in print. Accuracy was one worry, but there were also philosophical objections. Playfair was cavalier in his attitude to such concerns, his brashness and disregard for the opinions of others overriding all. There was a money-grubbing, opportunistic, and reckless aspect to Playfair's character—the attempted extortion was the work of a brazen and guileful, but also imprudent, individual. These disagreeable attributes, ironically, played midwife to the invention of statistical charts.

Sadly, the dark side of Playfair's character may have hindered acceptance of his graphs for nearly a century. It is not easy to dissociate the inventor from the invention—reputation and acceptance are closely bound, as much today as they were in the eighteenth century. Although unaware of the extent of Playfair's roguery, H. G. Funkhouser has also suggested that Playfair's work may have been slow in taking hold because of the inventor's unconventional opinions or behavior: "In damning the man, they may have damned his work as well. There is no evidence that the English scientists of his day recognized him as the creator of a method of representation and analysis that would become a universal language a century and a half later."[3]

4 Scaling the Heights (and Widths)

Playfair used graphs to inform and to convince. His understanding of the capacity of this new medium was remarkably deep. To gain a full appreciation for his skill, it is helpful to compare the polished displays he produced with the more rudimentary ones of his predecessors (a sampling of those appeared in chapter 1). We need not confine our comparisons to his predecessors, however, for Playfair's work does not suffer when placed side by side with those of more modern graphers.

Let us consider a simple table (table 4.1) that compares wages with living expenses over the 250-year span from 1570 (during the time of Elizabeth I) until 1820 (and the reign of George IV). Even a careful examination of this table provides little insight and excites even less emotion. But now let us look at what Playfair did with these long strings of dry numbers (in figure 4.1). The character of long-term trends in both the price of an important staple and the income of a representative trade are clear. The gap between the two is what draws our eye, as does the volatility of the price of wheat in comparison to the stability of wages. The message jumps out at us: some years it is much more difficult to afford bread than others. In addition, the steep increase in the price of wheat at the end of the eighteenth century coincides with the American Revolution. Apparently, when England was paying for guns, working people could ill afford either butter or bread. In addition, by including the identity of the reigning monarch Playfair is providing grist for a very specific sort of causal inference. And the figure is beautiful; the data are displayed with precision and style. The

Table 4.1
Comparison of Wages and Expenses, 1570–1820

Year	Price of wheat (Shillings/quarter)	Weekly wages of a "good Mechanic" (Shillings)	Reigning Monarch
1570	45	5.1	Elizabeth
1580	49	5.4	Elizabeth
1590	47	5.6	Elizabeth
1600	27	5.9	Elizabeth
1610	33	6.1	James I
1620	35	6.4	James I
1630	45	6.6	Charles I
1640	39	6.9	Charles I
1650	41	7.0	Charles I
1660	46	7.4	Cromwell
1670	38	7.8	Charles II
1680	35	8.2	Charles II
1690	40	8.6	James II
1700	30	9.0	William & Mary
1710	44	10.6	Anne
1720	29	11.7	Anne
1730	25	12.8	George I
1740	27	13.9	George II
1750	31	15.0	George II
1760	31	17.0	George II
1770	48	19.0	George III
1780	46	22.0	George III
1790	48	26.0	George III
1800	79	29.0	George III
1810	99	30.0	George III
1820	54	30.0	George IV

contrast with Robert Plot's barometric pressure plots (figure 1.5) is striking.

Playfair considered the same topic in different forms many times, but an early display from the first edition of his *Atlas* is relevant to the topic of this chapter. In figure 4.2, he plotted the interest on England's national debt. This figure has two aspects of special interest: first, the scale he chose, a topic to which we shall return shortly, and second, his choice of vertical grid lines. The data span the century from 1685 to 1785, yet the vertical grid lines are spaced irregularly: 1701, 1714, 1727, 1739, 1748, 1766, and so on. Grid lines are inserted at interesting points in time. We note that there is an increase in interest beginning in 1776. Indeed, if we look carefully we see that the interest on the national debt increases with every war. The tie-in to the increase in the price of wheat at about the same time is inescapable.

The facts are clear from figure 4.2, but Playfair understood that effective communication needs more than truth and clarity; it also needs impact. One of Playfair's methods for adding impact was the manipulation of the aspect ratio of the plot, the relationship between the scales of the horizontal and vertical axes. How to do this effectively is the subject of this chapter.

I have been a frustrated subscriber to the *New Yorker* magazine for decades. My basement bookshelves sag dangerously under the weight of back issues, saved only until I find the time to finish reading them. Sadly, their weekly arrival does not correspond to their monthly length. Although I seem never able to finish all the worthy articles, I do read all the cartoons. It is remarkable how often those cartoons refer to statistical graphics.

The Porges cartoon[1] on the next page serves

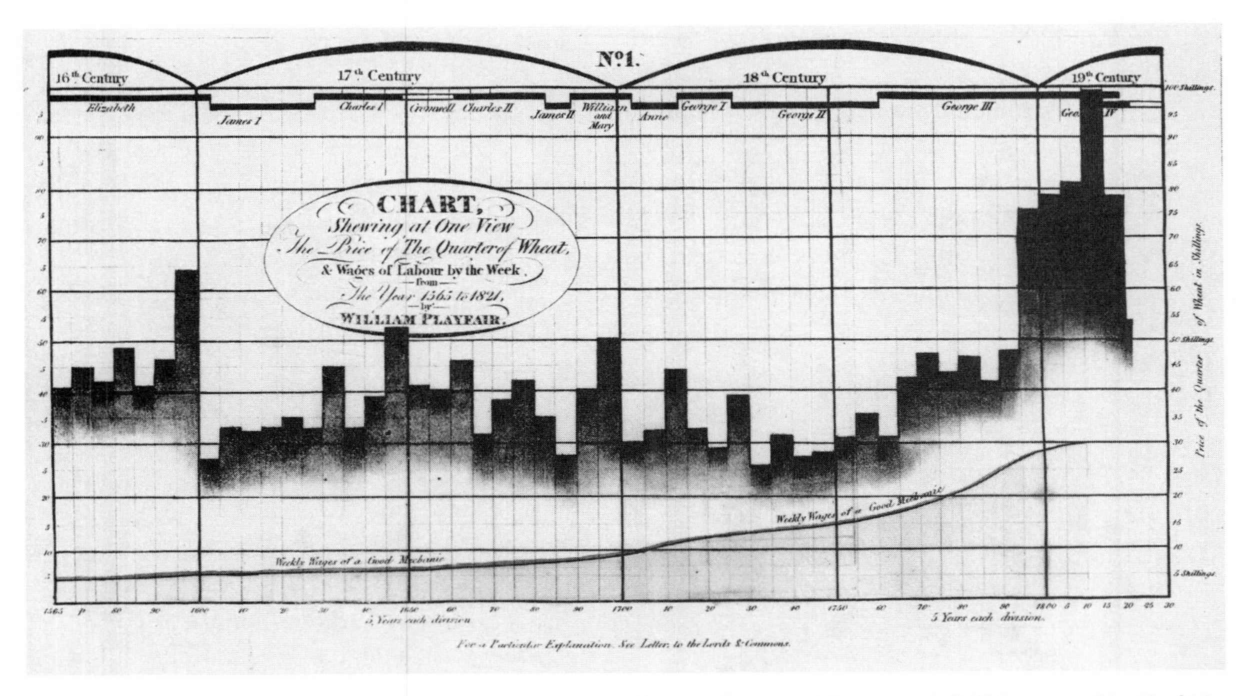

Figure 4.1 A figure containing three time series: prices, wages, and monarchs over a 250-year period. This appeared in Playfair's *Letter on Our Agricultural Distresses, Their Causes and Remedies; Accompanied with Tables and Copper-Plate Charts Shewing and Comparing the Prices of Wheat, Bread and Labour, from 1565 to 1821.* Princeton University Library, Department of Rare Books and Special Collections.

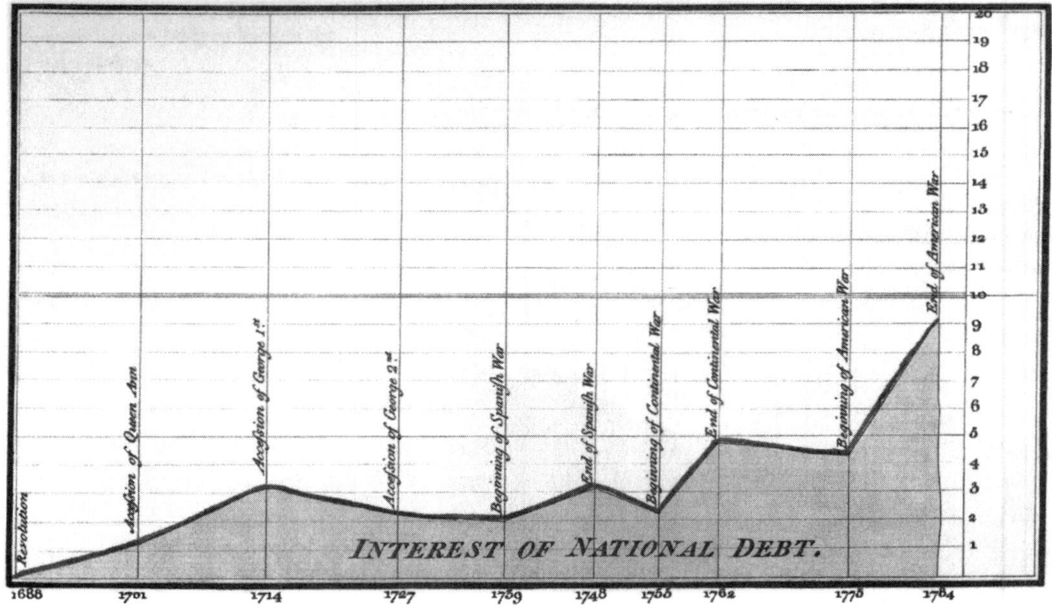

Figure 4.2 A display of the interest on England's national debt, in constant currency, from 1688 until 1784. The original of this figure was in the 1786 edition of Playfair's *Commercial and Political Atlas*. This version is reproduced from Edward Tufte, *Visual Display of Quantitative Information* (2d edition), p. 65, with permission.

as a vivid reminder that a graph is nothing but a visual metaphor. To be truthful, it must correspond closely to the phenomena it depicts; longer bars or bigger pie slices must correspond to more, a rising line must correspond to an increasing amount. If a graphical depiction of data does not faithfully follow this principle, it is almost sure to be misleading. But the metaphoric attachment of a graphic goes farther than this. The character of the depiction is a necessary and sufficient condition for the character of the data. When the data change, so too must their depiction; but when the depiction changes very little, we assume that the data, likewise, are relatively unchanging. If this convention is not followed, we are usually misled.

Consider the message conveyed by the simple bar graph in figure 4.3. The message that this graph conveys is that there has been a substantial decline in the number of elementary schools over the forty years depicted, but the number of private elementary schools has remained relatively constant.

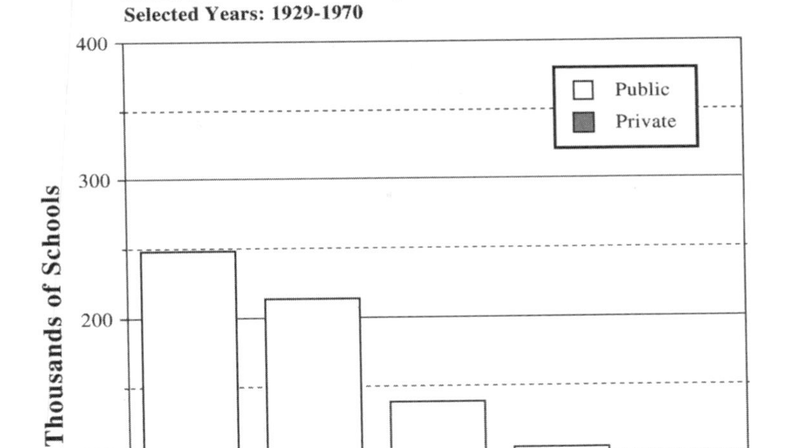

Figure 4.3. A fairly close copy of figure 6/4 from U.S. Bureau of the Census (1980), showing the decrease in the number of elementary schools over the forty years from 1930 until 1970.

The perception that the number of private elementary schools stayed about the same is false, because the scale chosen for the graph is too large to show the changes that did occur. If we isolate just the private schools and blow up the scale to make the changes visible, a very different story emerges (see figure 4.4). We see a profound increase in the number of private elementary schools between 1950 and 1960.

Why? One possible explanation is that the 1954 *Brown v. Topeka Board of Education* court decision, which declared segregated public schools illegal, instigated the founding of many private schools.

This explanation is supported if we plot more of the same data (figure 4.5), inserting biannual data points from 1929 until 1977. In this graph we see the number of private elementary schools meandering along at about ten thousand until the start of the era of school integration, when the number of schools increased more than 50 percent. The decade of shenanigans whose goal it was to try to subvert the spirit of *Brown* seems to come to a crashing end with the passage of the 1964 Civil Rights Act. This explanation may be all wet, but without viewing the data in a proper scale, even the need for explanation was not

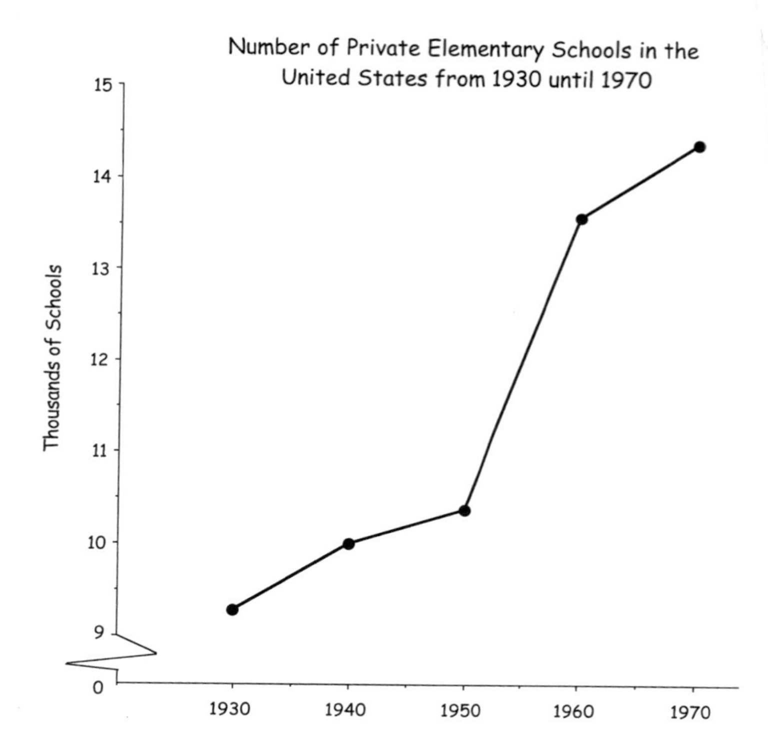

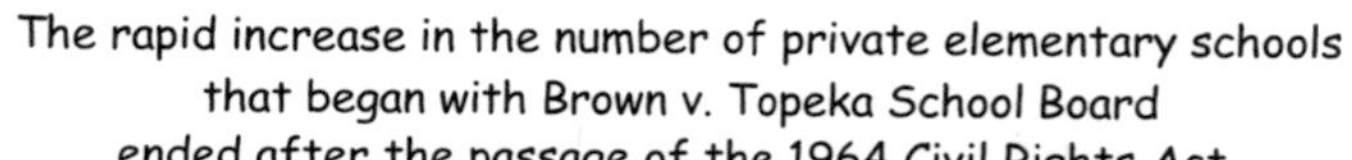

Figure 4.4. Expanding the scale and plotting just the private elementary schools reveals a sharp increase in their number during the decade of the 1950s.

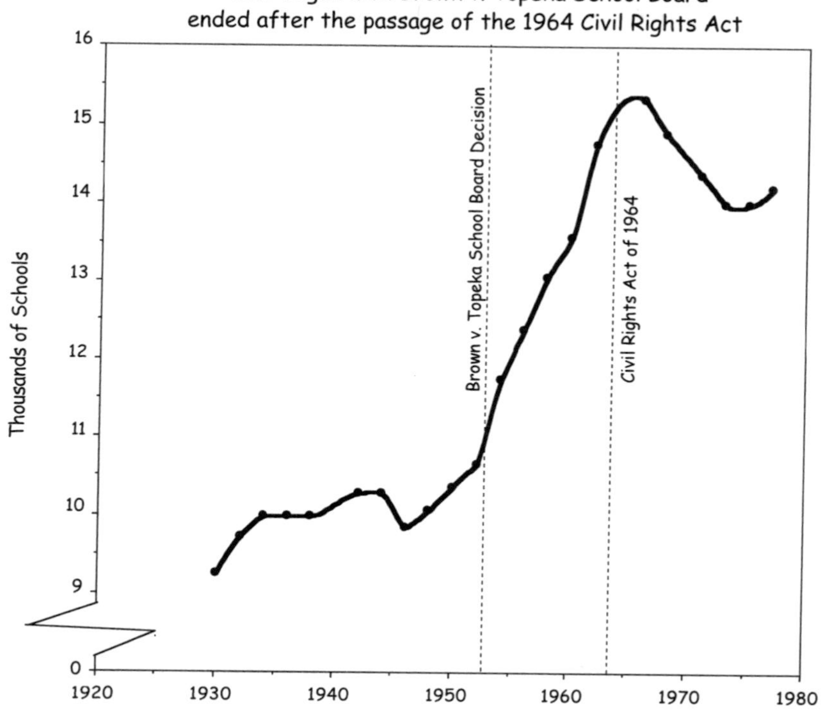

Figure 4.5. Adding more data points to figure 4.4 and augmenting the plot with two important events suggests a plausible causal inference (but see note on facing page).

apparent.* Thus the original chart violates the fundamental tenet of honest display, which requires that the display be a metaphor for the data—if it looks as though nothing is going on, then nothing should be going on.

But what is the proper scale? This is a difficult and subtle question. The appearance, and hence the perception, of any statistical graphic is massively influenced by the choice of scale. If the scale of the vertical axis is too narrow relative to the scale of the horizontal axis, random meanders look meaningful.

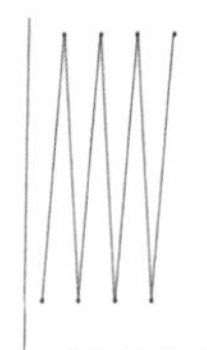

Too wide a vertical scale (as in figure 4.3) can make potentially important variations all but disappear.

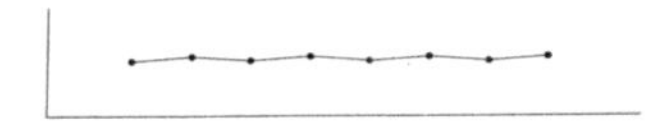

Is there a rule that might guarantee a wise choice?

In 1914, Willard C. Brinton recognized this issue but, other than urging care, offered no solution. He wrote: "The beginner in curve plotting and in curve reading is apt to be somewhat puzzled by the different effects which may be obtained by changing the ratio between the vertical scale and the horizontal scale. . . . Just as the written or spoken English language may be used to make gross exaggerations, so charts and especially curves may convey exaggerations unless the person uses as much care as he would ordinarily use to avoid exaggeration if presenting his material by written or spoken words."[2]

Seventy years later, Calvin Schmid made the similar observation that "grid proportions are of pronounced significance as the determinants of the visual impression conveyed. . . . If the vertical dimension were greatly elongated, the configuration of the curves would be characterized by sharp and steep trends and fluctuations. . . . [I]f it is made extremely wide the corresponding curves would tend to become flat."[3] But when it came to providing guidance on how to determine the correct scale, Schmid fell short, telling us only, "The scales chosen for both axes should result in a well proportioned chart . . . [that] should not minimize or exaggerate variations in the curves" (p. 19).

Even though Schmid did not provide an

* We are well aware of the dangers inherent in trying to make causal inferences from such flimsy evidence, no matter how suggestive. Take, for example, a figure showing national average SAT scores from 1960 to 2000. It is well known that these scores peaked in the mid-1960s before a steep, decade-long decline that bottomed out around 1980. Many explanations have been offered for this decline. The one that has the greatest support in the educational community is that the democratization of education that accelerated in the 1960s increased the size of the college-going population enormously and hence lowered the average score. This explanation fully accounts for the decline until 1972. But these demographic shifts were essentially complete by 1972, and yet the decline continued unabated. A Christian fundamentalist newspaper produced a version of the accompanying graph with the Supreme Court school prayer decision line as a guide to causal inference. I added the second line and made the moral of the graphical story explicit.

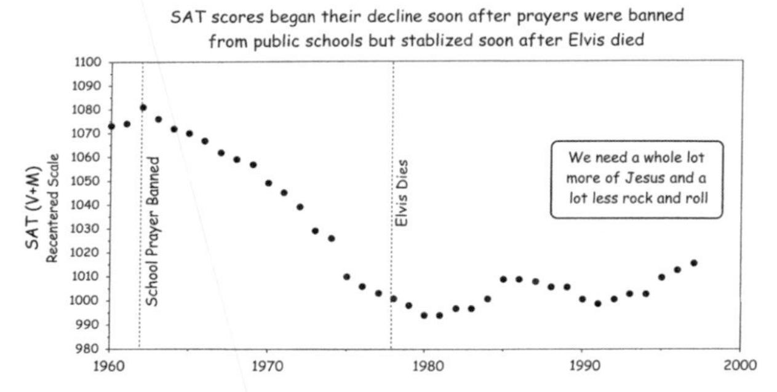

A plot of the national average Scholastic Aptitude Test (SAT) score (verbal + mathematical) from 1960 until 1997, with two possibly related events indicated and one causal inference suggested

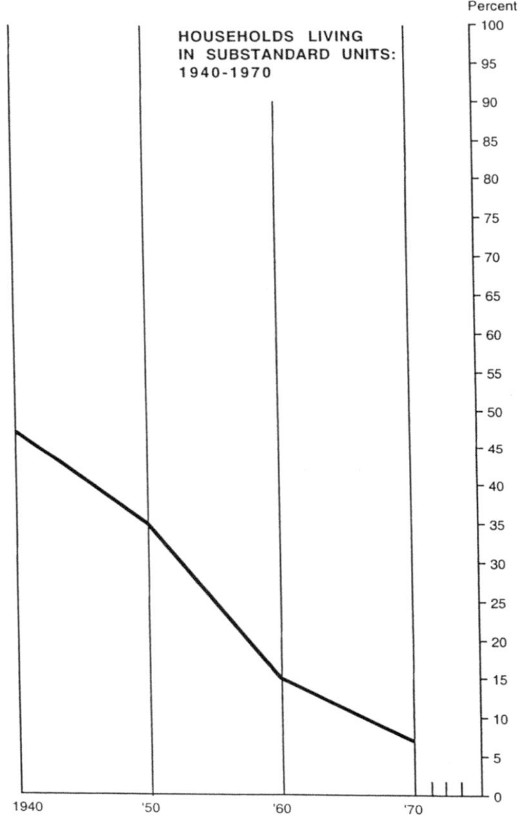

Figure 4.6. An accurate rendering of a graph from U.S. Office of Management and Budget (1973, p. 190). Schmid criticized this on four separate counts, but his principal concern was the scale running wastefully well beyond the data.

operational rule to help us make wise decisions about scale choice, he did profusely illustrate versions of bad practice, along with a running commentary on their flaws. For example, he reproduced a plot from a publication of the Office of Management and Budget (see figure 4.6), and the first of his criticisms is that "half the grid in the vertical dimension is superfluous" (p. 22). Thus, Schmid's stand on improper proportioning of scales in graphs is akin to Justice Potter Stewart's on pornography: he may not be able to define it, but he sure knows it when he sees it!

The difficulty that graphical scholars have in defining proper proportions is understandable. Those involved in data detective work tend to want to use the smallest scale that allows the inclusion of all the data within the plot frame. People interested in only the "big picture" may prefer a scale selection that diminishes the visual impact of minor jiggles and keeps the overall scale clearly in view by including markers such as a zero line. Each point of view has logical supporting arguments.

The distinguished statistician Bill Cleveland, not surprisingly, takes the detective's point of view: "Scale issues are subtle and difficult. . . . [One should] choose the scales so that the data rectangle fills up as much of the scale-line rectangle as possible."[4] He then goes on to amplify this advice in several wise ways, including the directive that when the goal of the graph is to display change, the ratio of the vertical and horizontal axes (the aspect ratio) should be chosen so that the slope of the data is about 45 degrees. Cleveland's contention is that blindly following conventions such as "always including zero" may lead you to have an overlarge scale, which in turn may hide important aspects of the data. I have great sympathy with his conclusions, as they not only make intuitive sense but also are based on strong empirical and theoretical foundations. One can always replot a too-detailed picture and thus get the smoothed view, but once the details are lost, the retrieval task is often impossible.

Though Cleveland's approach may not be a definitive answer to scaling graphs, is it a good working default option? In investigating the value of this advice, I looked at some of the graphical works of the masters. For the most part, the graphics of Etienne Charles Marey and Charles Joseph Minard follow Cleveland's advice. So too does Playfair, but with some exceptions. In this instance, the exceptions are instructive—instructive both about graphical construction and about the workings of the mind of William Playfair.

The structure of Playfair's *Atlas* is very regular. Each chapter typically begins with a graph and then discusses the content of that graph and its implications for the revenues, expenditures, debts, and commerce of England. Most of the forty graphs more or less abide by Cleveland's general rule. For example, figure 4.7, a chart of England's national debt from the *Atlas*'s third edition, starts at zero and extends to five hundred million pounds. This range is about 20 percent larger than the data, but in being so it allows room for the elegant title and the increased debt that seemed certain to accrue in future years.* Its slope seems awfully close to Cleveland's 45 degrees. Thus I would conclude that Playfair would agree with subsequent graphic scholars about using the full size of the graph to just cover the data and in their choice of aspect ratio.

This display is worthy of deeper study for several reasons. Compare the impact of the soaring debt shown on this scale with the more modest impression made by the growing interest in his earlier display (figure 4.2). In the fifteen years between the first edition of his *Atlas* and the third (which contained this display), did Playfair learn something about choice of scale and visual impact? Or did the relationship between the stability of society and the cost of bread become clearer to him after his experiences during the French Revolution? (See chapter 2.) Surely anyone seeing Playfair's plot of national debt comes away with a strong causal connection between debt and war (and, from figure 4.1, the price of bread). Whatever the cause, by 1801 Playfair was not a disinterested observer and wanted to provide a compelling argument for his point of view. Although we can only speculate on why he changed scale in the third edition of his *Atlas*, there is evidence, which we shall discuss next, that suggests that he understood how to choose a scale for maximum effect in 1786, but instead followed another route.

Most of the *Atlas*'s chapters deal with the monetary value of the trade between England and another nation over the eighteenth century. They typically follow an identical format: a graph and then several pages of discussion of the graph and its implications. The graph's content is a line that depicts the value of England's exports to that nation, and a second line corresponding to the value of the goods that England imports from that nation, with the space between those two lines being shaded and labeled as the balance in favor or against England. The time (horizontal) axis usually ranges from 1700 to 1800, and the

* This graph has a number of features of interest. The most important one for contemporary designers is the unevenly spaced grid. Using the default options of most statistical graphic software, the data from which this chart is composed would generate tick marks every ten years starting from 1680 and ending at 1810. But do we care about 1700 or 1770 or 1780? Playfair's genius was in the placing of grid lines at points of interest based on the data and labeling them. In this way answers are provided for the natural questions that anyone looking at the chart would ask.

Question: When did the debt begin to skyrocket?

Answer: 1730, 1755, 1775, and 1793.

Question: Did those years share any common event that would have an untoward effect on England's debt?

Answer: Each of those years corresponds to the start of a war, for example, 1730 to the start of the Spanish War; 1755, the Seven Years War; and 1775, the American Revolution.

Question: When did the debt begin to subside?

Answer: 1748, 1762, and 1784.

Question: Was there any common event among those dates?

Answer: Each of those years corresponds to the end of a war.

Question: Is there any general conclusion you can draw from this chart?

Answer: Wars are bad for the national debt.

Of course, the skyrocketing debt is somewhat exaggerated since the figures are not adjusted for inflation. Playfair knew about making such adjustments: a few pages later (p. 129) he presented an adjusted plot of the interest on the national debt, so we must assume that he chose not to adjust in order to emphasize the phenomenon. For a further discussion of this, see Tufte 1983, pp. 64–65.

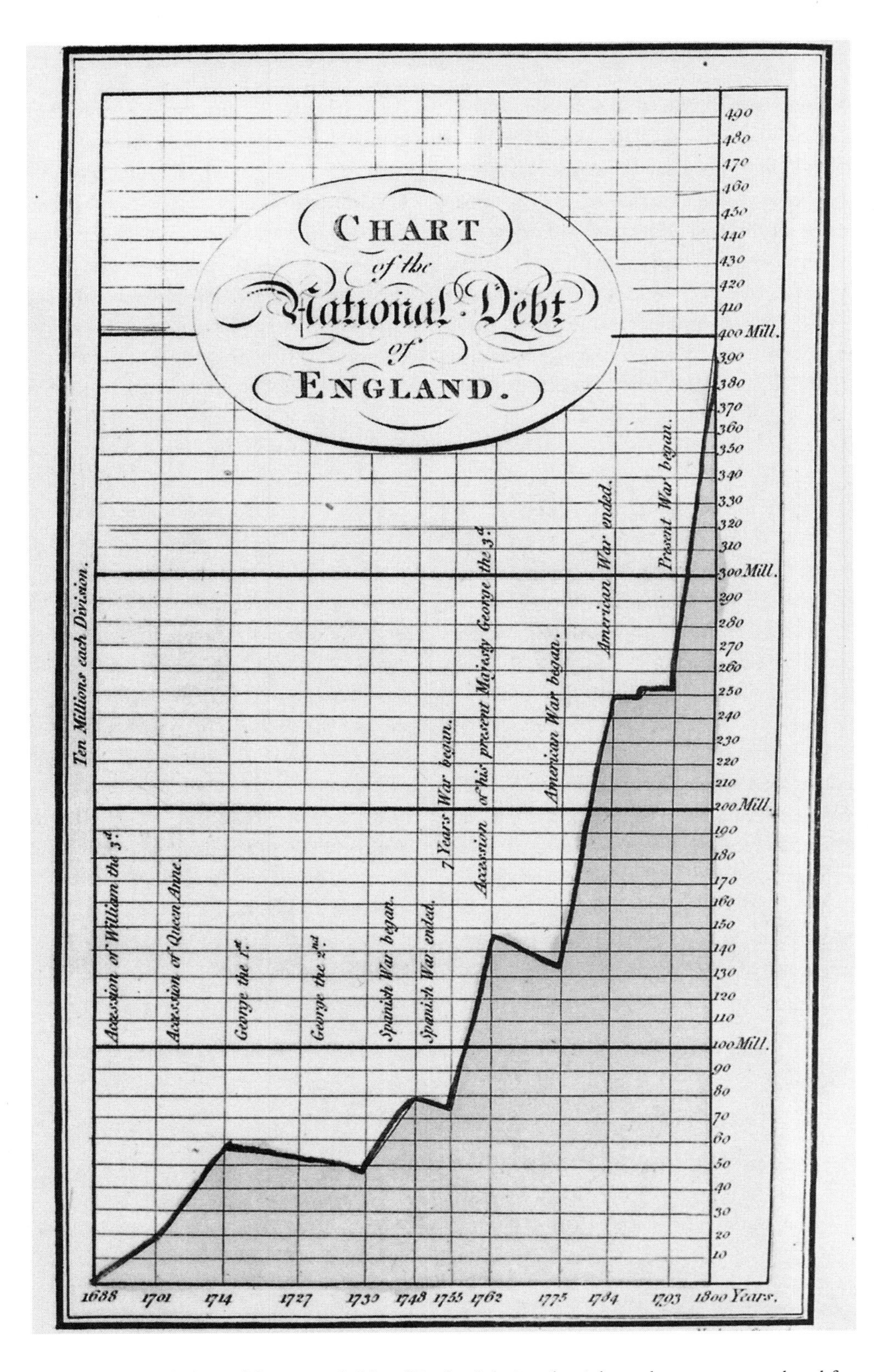

Figure 4.7. An annotated chart of the national debt of England during the eighteenth century, reproduced from the third edition of Playfair's *Commercial and Political Atlas* (1801). Reprinted with permission of the Princeton University Libraries from a copy of this work that is held in the Orlando F. Weber Collection of Economic History. Princeton University Library, Department of Rare Books and Special Collections.

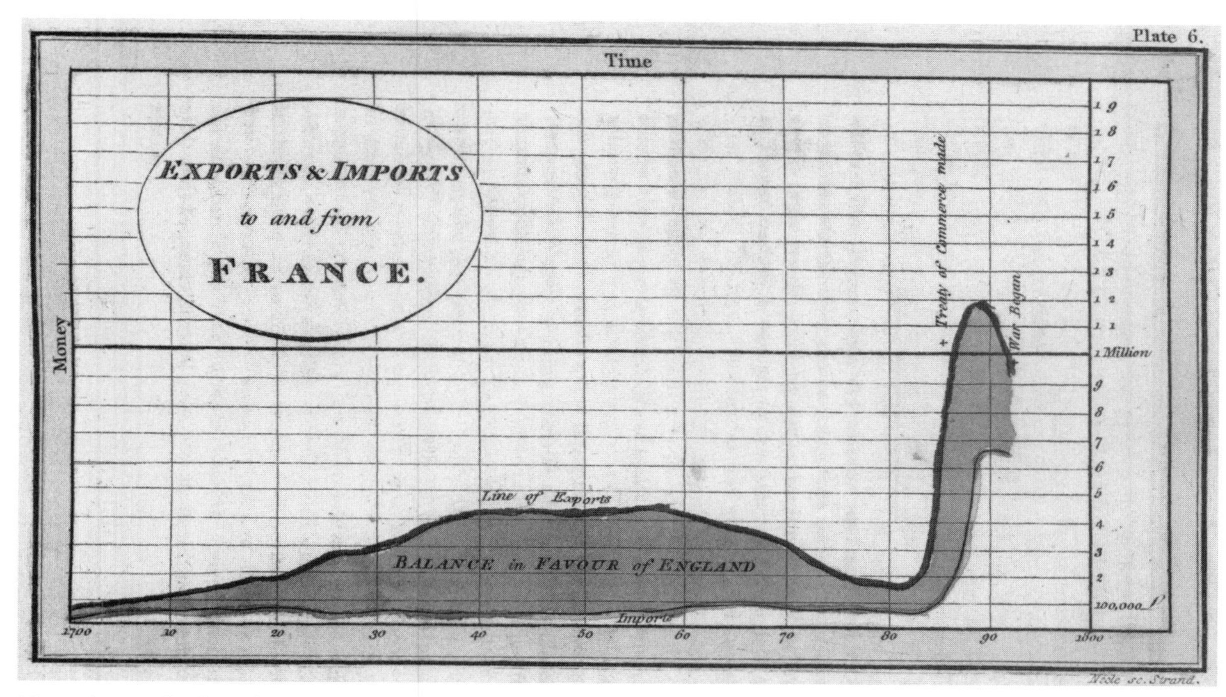

Figure 4.8. England's trade with France between 1700 and 1792, from the third edition of Playfair's *Atlas* (1801). Reprinted with permission of the University of Pennsylvania Libraries from a copy of this work that is held in the Van Pelt–Dietrich Library Center, Annenberg Rare Book and Manuscript Library.

money (vertical) axis ranges from zero to a little beyond the maximum amount of trade for that country. But one exception stands in stark contrast to the advice about scale that Playfair seems to follow in his other plots—the trade between England and France.

In figure 4.8, the greatest data point is about £300,000, but the scale goes up to one million. Why did Playfair squish the France-England data down into the bottom of the plot? The impression we get from the plot is not one of robust trade between two important partners. Rather it looks as if hardly any trade at all is going on. This graph would draw the attention of knowledgeable readers, who would immediately spot the disparity between what they know the trade must be and the impression carried by the graph. Why would Playfair do this? Didn't he know that by choosing an overlarge scale he would convey the wrong impression? Did the master make a mistake?

The answers to these questions leap from Playfair's discussion on the pages following this figure. He wrote,

> We have now before us a very fallacious representation of the trade between two countries, which, from their situation, as well as from the nature of their productions, we might expect to find immense; yet which, through a strange species of policy, is extremely inconsiderable.
>
> There cannot be a doubt that the illicit trade far exceeds in amount that here delineated, which can include only what is regularly entered. This trade furnishes us with an astonishing instance of the inefficiency of the laws that are injudiciously enacted, and which furnish too great a reward for evasion.

Playfair wasn't confused. He knew exactly what he was doing. He thought that the large import duties were not affecting trade overall, just the legal (hence recorded) trade, and he used the scale to convey a truthful impression

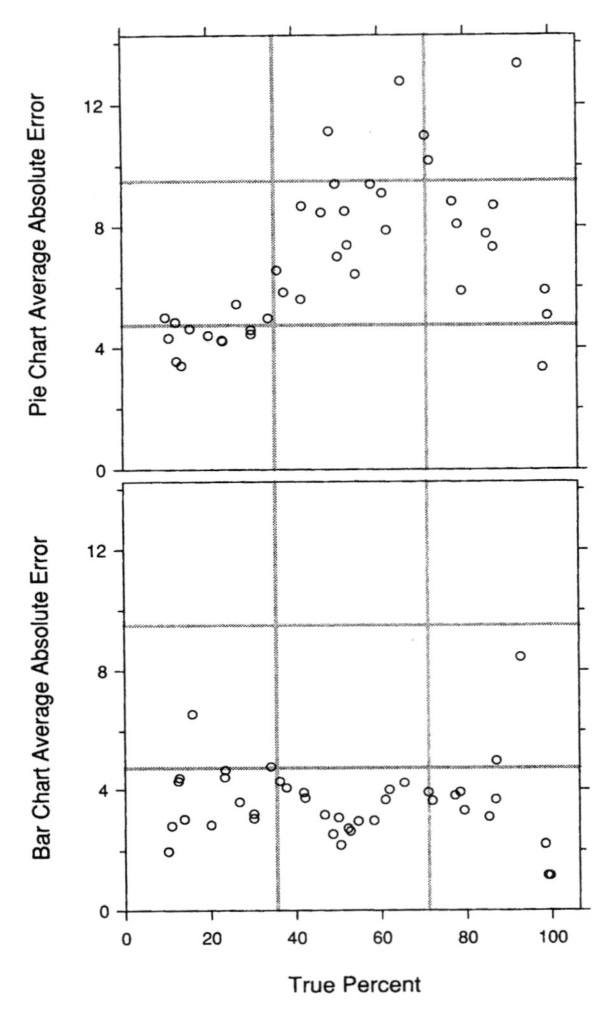

Figure 4.9. Scales on both panels are identical, thus facilitating comparisons between them. This is figure 2.54 from Cleveland (1994, p. 87), and is reprinted with permission.

from incorrect data. But what does this mean about Cleveland's rule about choosing "the scales so that the data rectangle fills up as much of the scale-line rectangle as possible?" Is Playfair's use of the scale an exception that Cleveland hadn't thought of? No, for shortly after providing this advice Cleveland offers an amendment (p. 86): "Choose appropriate scales when data on different panels are compared."

He illustrates this, appropriately enough, with data from an experiment on perceptual error when viewing two different formats of statistical graphics (figure 4.9). Note that the data from the bar chart fall in the bottom of the scale range required to include the data points from the pie chart. Yet this considerably eases the viewer's task, and we can tell immediately which format is superior for the task tested (it would have been easier still had the two panels been placed side by side instead of top to bottom, but that is a minor point).

The only difference between what Playfair did and what Cleveland recommended is that Playfair's comparison was implicit. He was comparing the recorded trade between the two countries to what informed intuition would have suggested. It is interesting that in the update of this chart in the third edition of his *Atlas*, Playfair showed a huge increase in the trade with France, roughly corresponding to the time that the 1785 Treaty of Commerce was made. But apparently he felt that it was still too little, for he doubled the scale (up to two million pounds) and appended the comment, "that treaty was not on good principles; and [it was approved] during a time in which the French government was going to ruin."

We conclude with happy news. The difficult task of properly setting the scale of a graph remains difficult but not mysterious. There is agreement among experts spanning two hundred years. The default option should be to choose a scale that fills the plot with data. We can deviate from this under circumstances when it is better not to fill the plot with data, but those circumstances are usually clear. It is important to remember that the sin of using too small a scale is venial; the viewer can correct it. The sin of using too large a scale cannot be corrected without access to the original data; it can be mortal.

5 A Priestley View of International Currency Exchanges

As middle age and its associated onrushing senescence swoop down on me, I have come to appreciate the wisdom of the Talmudic *midrash* attributed to Rabban Yohanan ben Zakkai: "If there be a plant in your hand when they say to you, Behold the Messiah!, go and plant the plant, and afterward go out to greet him." The scientific establishment regularly discovers new sources of redemption—this year's promise is that computer-aided design will save us all—but I suspect that the Lord is best served by scientists who find salvation in the routine transactions of their daily work.

The economist Michael Melvin produced a graph showing the operating hours of nine major currency exchanges (see figure 5.1). The currency exchanges are represented by an entangled mesh of elongated boxes whose ends bend over and discontinuously stretch to a time scale at the bottom. The length of the horizontal box is proportional to the amount of time they are open. The lines are superimposed over a world map, although the relationship between the geographic placement of the lines and usable information eludes me. This is an example of using computer-based graphics to complicate the simple.

What questions might this graphic be meant to answer? There are some obvious ones such as, "When is the New York currency exchange open?" or "Where can I trade now?" Or more subtle questions such as, "Which exchanges are open for the longest periods of time?" or "the shortest?" Or, finally, "Is there any time in the day or night when all of these exchanges are closed? When?" The information to answer all of these questions is contained in figure 5.1, but it is not easy to extract. And this is with only nine exchanges. Suppose there were twenty? Or thirty? Or fifty? This design would be hopeless.

How can it be improved? One obvious redesign is shown in figure 5.2. All the questions phrased above can be answered trivially. The Singapore exchange, which is open for nearly eleven hours a day, narrowly beats out the Hong Kong exchange, and they both dwarf the lazy American currency traders, who seem to think that seven hours constitutes a full workday. We are struck with horror that, after the San Francisco exchange closes at 11:00 p.m. (GMT), there is an entire hour of trading quietude until the Hong Kong exchange opens.

The ease of interpretation available in figure 5.2 comes from omitting all unrelated figurations

I am grateful to Martin Gilchrist for the example from Melvin's book that prompted this essay. I was first made aware of Priestley's graph in a 1990 article by Albert Biderman. Parts of the introduction and conclusion are paraphrased from some remarks made by Princeton's Marvin Bressler and reproduced in the November 15, 1992, issue of the *Princeton Alumni Weekly*. I remain in awe of his erudition.

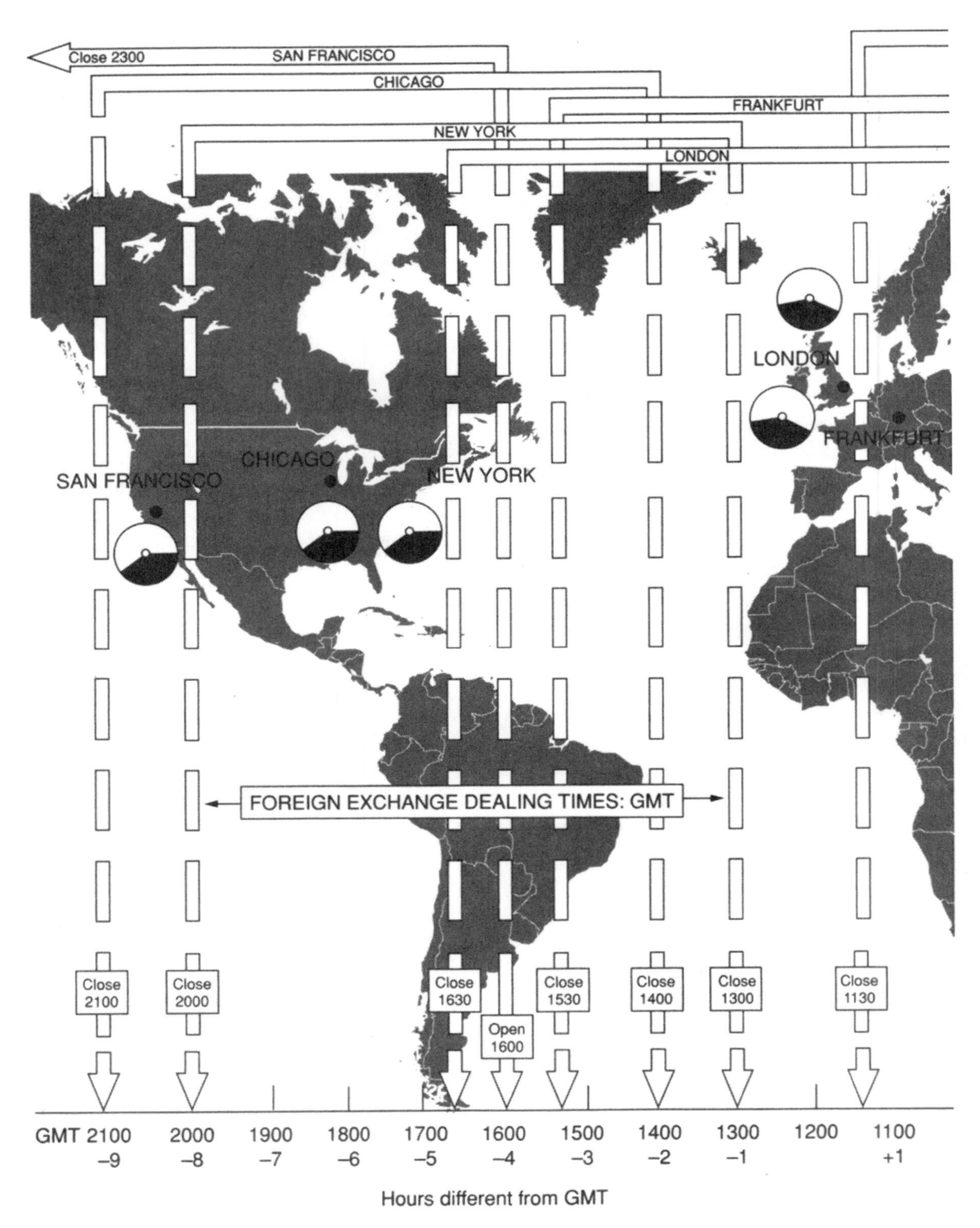

Figure 5.1. Operating hours of nine major currency exchanges (Melvin 1995, pp. 10–11) shown on a geographic background.

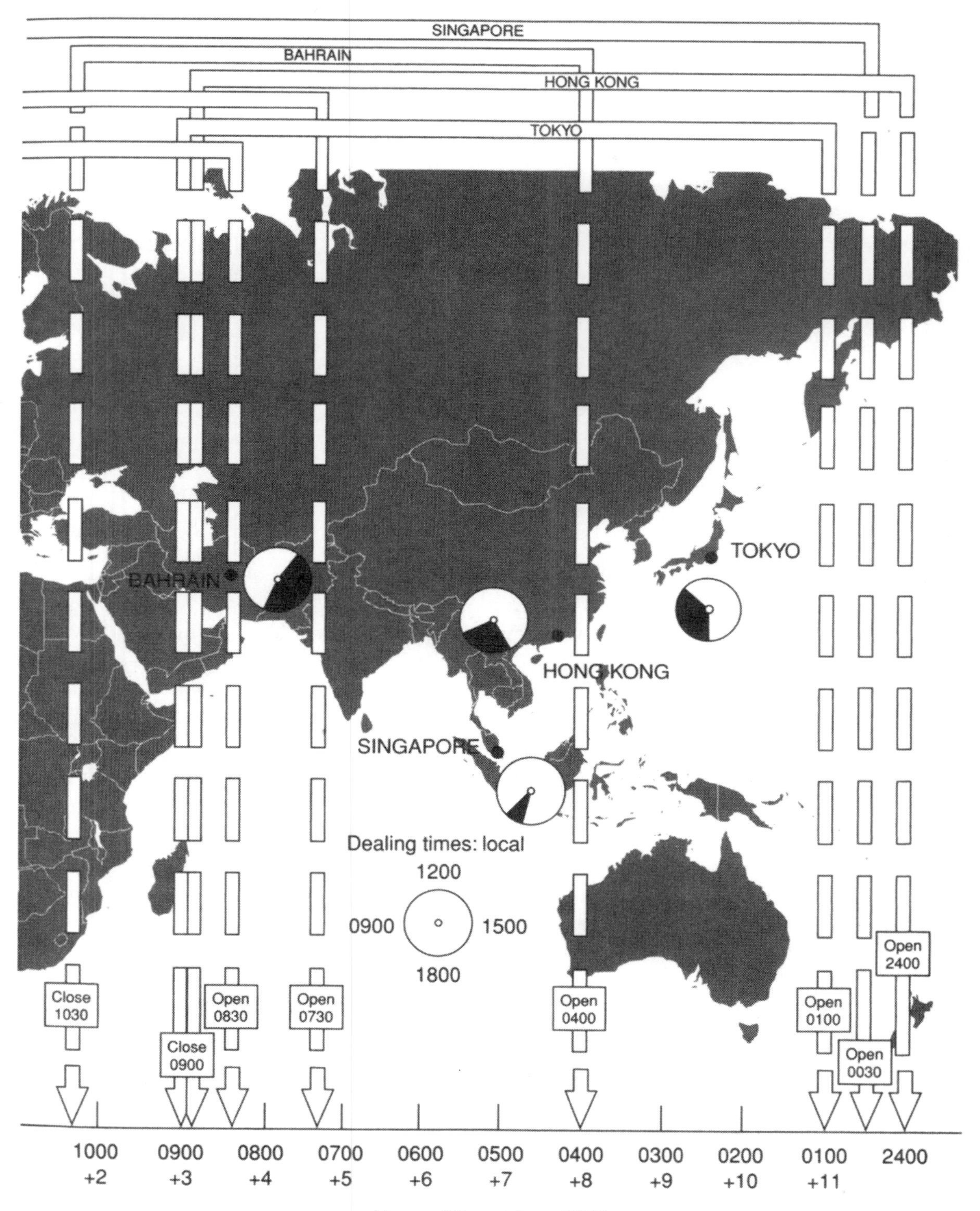

SINGAPORE
BAHRAIN
HONG KONG
TOKYO
TOKYO
BAHRAIN
HONG KONG
SINGAPORE
Dealing times: local
1200
0900
1500
1800
Close 1030
Close 0900
Open 0830
Open 0730
Open 0400
Open 0100
Open 0030
Open 2400
1000 +2
0900 +3
0800 +4
0700 +5
0600 +6
0500 +7
0400 +8
0300 +9
0200 +10
0100 +11
2400
Hours different from GMT

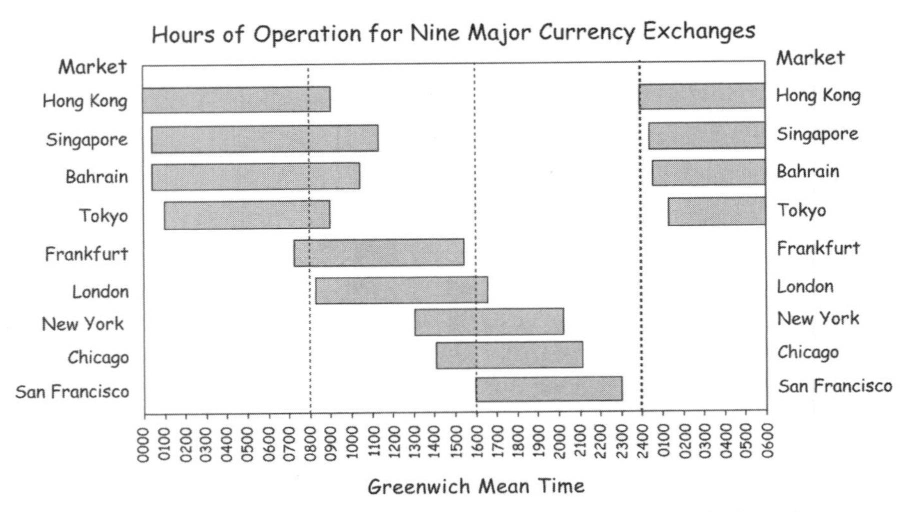

Figure 5.2. Foreign currency exchange dealing times shown clearly without chartjunk.

(what Tufte calls "chartjunk"). But why were they put in there in the first place? One cannot be sure, but it is discouraging to believe that anyone would actually work hard to fill up a graph with a bunch of stuff that effectively hides the message of the data. Thus the only sensible conclusion is that this was done easily via a computer graphics program that allows the trivial superposition of anything over anything.

What makes this entire exercise of special interest is that the graphical form shown in figure 5.2 is very old indeed. In 1786, William Playfair, lacking the longitudinal data for Scotland, needed to show the same sorts of time trends that he displayed for other countries, used a horizontal bar chart to show the imports and exports of Scotland with eighteen trading partners (see figure 6.1). But he did not invent the form. In fact, he gave credit to Priestley (1765; 1769), who plotted the life spans of fifty-nine famous people who lived in the six centuries before Christ (see figure 5.3).

Priestley's plot has several features of interest. He duplicated the scale on top and bottom, thus making it easier to read; he divided the fifty-nine people into two groups, thirty "Men of Learning" and twenty-nine "Statesmen"; he invented the convention of using dots to indicate uncertainty about the exact date of birth or death; and he used tick marks to demark the axis. It is interesting to note that the iconic representation of people's life spans with bars was not deemed obvious by Priestley, who filled several pages with explanation to justify this as a natural procedure. His graphs enjoyed immense popularity, and in a revision four years later (1769) he omitted all explanation.

Abetted by computer graphic software, it is easy to see why scientists who draw graphs add in extras like the world map in figure 5.1. Through the inclusion of a geographic context perhaps the author of this graph felt he was better able to give his audience a glimpse of the world. But, returning to the Talmudic spirit of the beginning of this chapter, in so doing he lost track of his primary, though perhaps more pedestrian goal, the clear communication of relevant information.

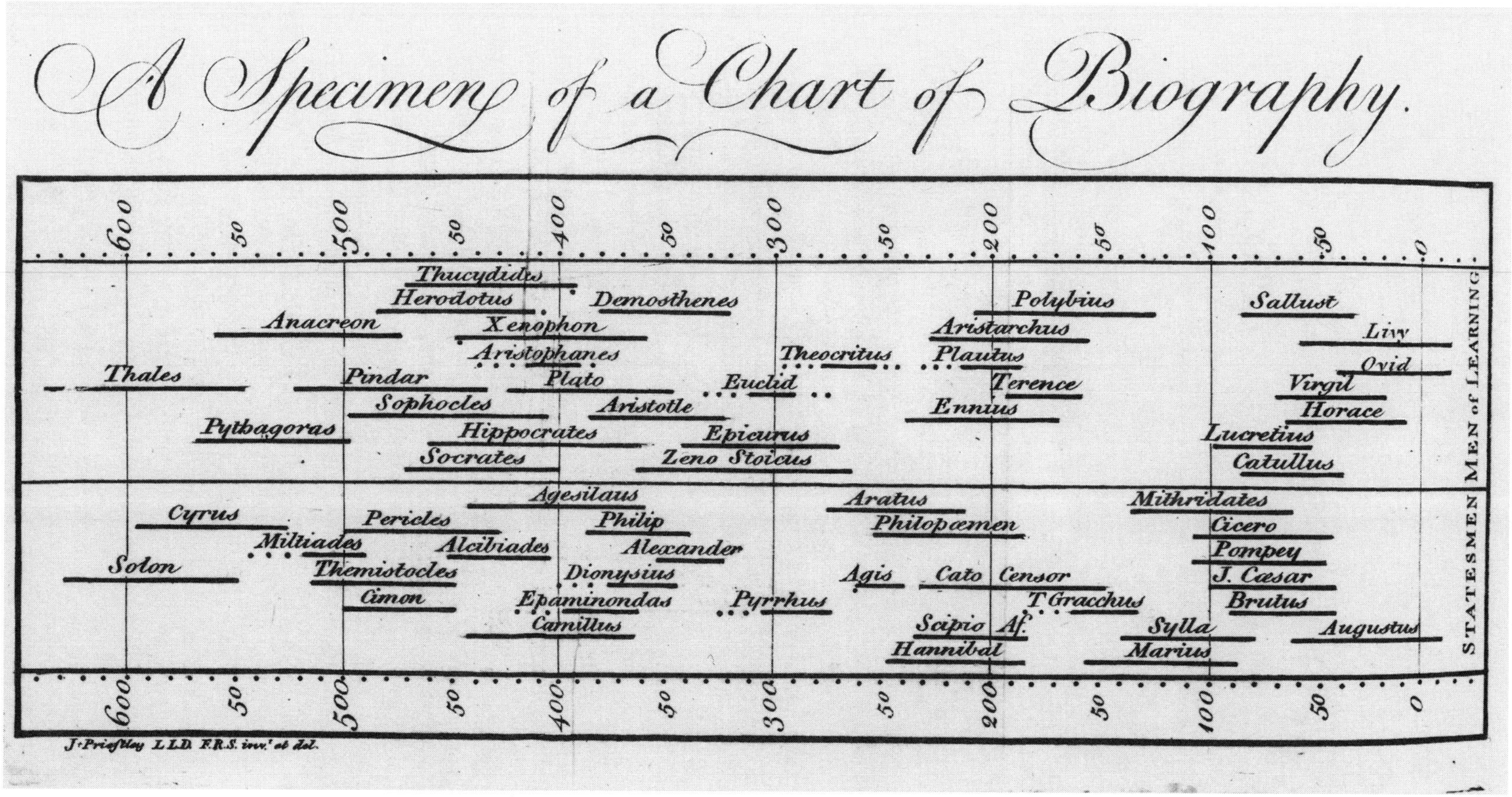

Figure 5.3. Lifespans of fifty-nine famous people in the six centuries before Christ (Priestley 1765). Princeton University Library, Department of Rare Books and Special Collections.

6 Tom's Veggies and the American Way

When Buck Mulligan began James Joyce's *Ulysses* by looking into the broken mirror of Irish art, he provided a powerful image of how the character of a society reflects itself in all forms of expression. The brash, revolutionary, democratic structure of America has shown itself in all forms. Tom Wolfe, in the lead essay of *The Kandy-Kolored Tangerine-Flake Streamline Baby*, maintained that there were only two cities in the world that had completely coherent architectural themes: Versailles and Las Vegas. Versailles was built in the classical style favored by eighteenth-century French aristocracy; Las Vegas in what Wolfe termed "MacDonald's Parabola Modern." The latter style is characterized by seven-story-high neon lights and an aesthetic in step with paintings of trucks on black velvet. Wolfe's point, germane to the topic of this chapter, was that although there has always been an enormous diversity of taste within all societies, historically it was only those with aristocratic tastes that had the resources to erect monuments to those tastes. Only in America were ordinary people with nonaristocratic tastes able to garner such resources—hence Las Vegas.

Everywhere we look we see manifestations of the American spirit. In TV programming, no one confuses the offerings of the BBC with Beavis and Butthead; the country that produced Ingmar Bergman could not have made Star Wars; Paul Bocuse's culinary creations are not at home with fries and a shake to go.

But if the enshrinement of popular taste were the principal contribution of America to the history of culture, it would hardly be worth a celebration. No, it is far more than that. America's democratic spirit allows an eclecticism that existed nowhere else. America accommodates it all. And this breadth of accommodation is not merely a reflection of a diverse population, because the spirit of this diversity can also invade each person: Steven Spielberg created both *Schindler's List* and *Indiana Jones*; Leonard Bernstein composed masses for concert halls and *West Side Story* for the masses; Richard Feynman played with the fundamental building blocks of the universe and with bongo drums.

In 1765, Joseph Priestley invented the line chart (see figure 5.3) and used it to depict the life spans of famous figures from antiquity; Pythagoras, Socrates, Pericles, Livy, Ovid, and Augustus all found their way onto Priestley's plot. Priestley's demonstration of this new tool was clearly in the classical tradition.

Twenty-one years later, William Playfair used a variant on the same form (see figure 6.1)

I am grateful to Albert Biderman for bringing Thomas Jefferson's wonderful graph to my attention. I would also like to thank Andy Baird for his help in electronically augmenting both Priestley's and Jefferson's plots, and Linda Steinberg for helpful comments on an earlier draft.

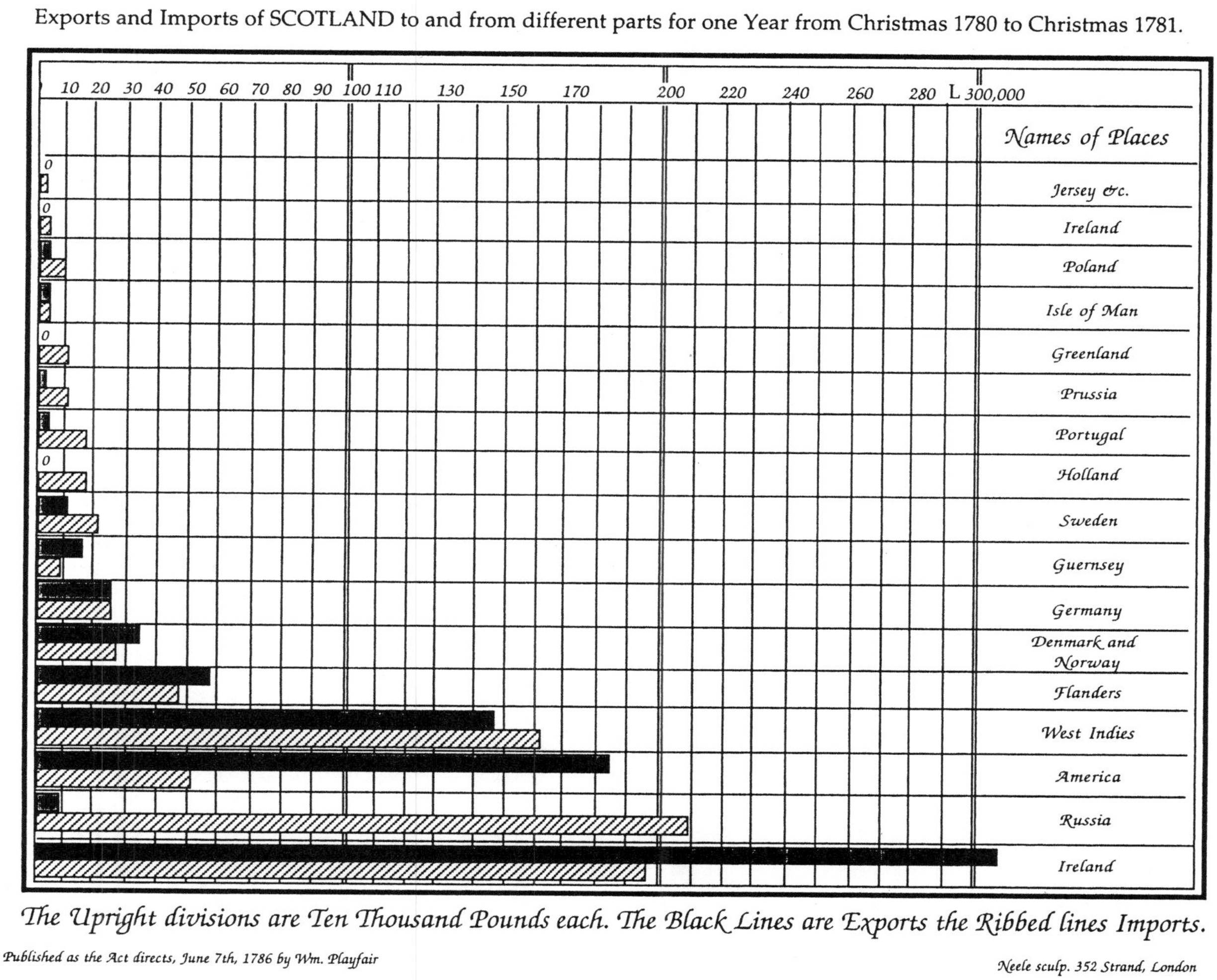
Exports and Imports of SCOTLAND to and from different parts for one Year from Christmas 1780 to Christmas 1781.
10 20 30 40 50 60 70 80 90 100 110 130 150 170 200 220 240 260 280 L 300,000
Names of Places
Jersey &c.
Ireland
Poland
Isle of Man
Greenland
Prussia
Portugal
Holland
Sweden
Guernsey
Germany
Denmark and Norway
Flanders
West Indies
America
Russia
Ireland
The Upright divisions are Ten Thousand Pounds each. The Black Lines are Exports the Ribbed lines Imports.
Published as the Act directs, June 7th, 1786 by Wm. Playfair
Neele sculp. 352 Strand, London

Figure 6.1. Scotland's exports and imports with eighteen trading partners (after Playfair 1786, plate 23).

to show the extent of imports and exports of Scotland to eighteen other places. Playfair, as has been amply documented (chapters 2 and 3), was an iconoclast, quite different from many of his scientific brethren in Europe. He used the line graph to show economic data. This plot format was not chosen because of any special affection for it, but rather of necessity, because he lacked the time-series data he needed to show what he wanted. In his own words, "The limits of this work do not admit of representing the trade of Scotland for a series of years, which, in order to understand the affairs of that country, would be necessary to do. Yet, though they cannot be represented at full length, it would be highly blameable intirely to omit the concerns of so considerable a portion of this kingdom."

Playfair's practical bent provides a sharp contrast to the classical content chosen by Priestley to illustrate his invention.* But what about the American aristocracy, such as it was?

* Playfair's long association during his formative years with great inventors made for a person attuned to the practical, rather than the theoretical; to coming up with anything that worked well for things that were important to daily affairs of men—business and the politics with which business was closely allied. The theories that concerned him most were the practical

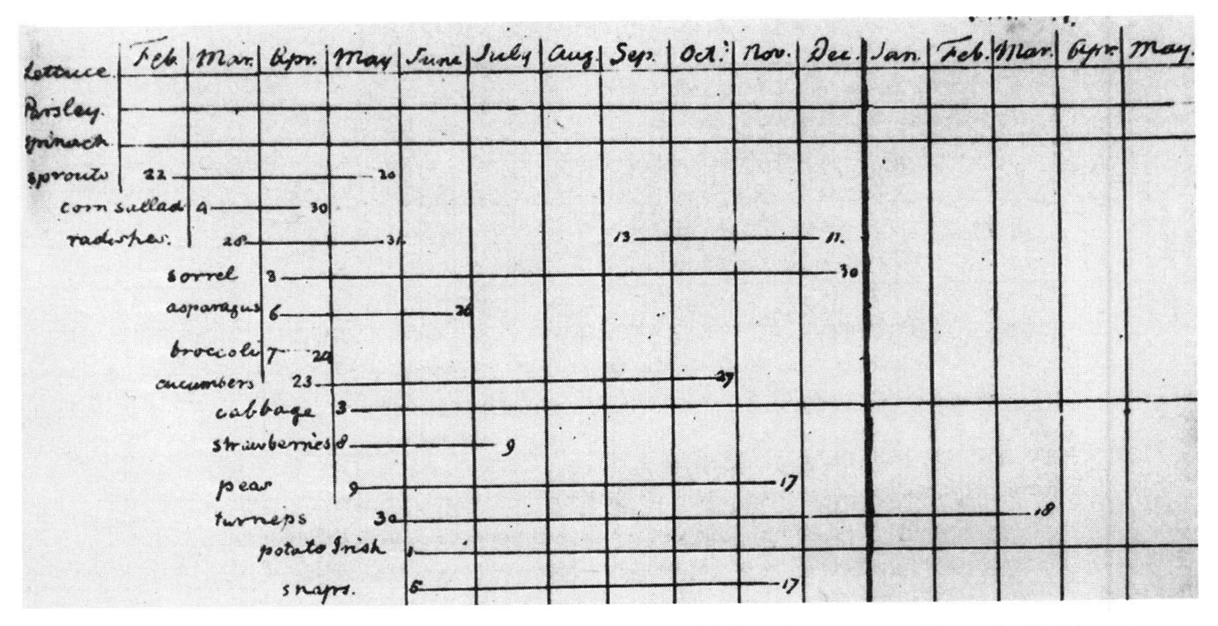

Figure 6.2. An excerpt from a plot by Thomas Jefferson showing the availability of sixteen vegetables in the Washington market during 1802. This figure is reproduced, with permission, from Froncek (1985, p. 101).

Thomas Jefferson spent five years as ambassador to France (from 1784 until 1789). During that time he might have come to know both Playfair personally, but he certainly became acquainted with his graphical inventions. Although Jefferson was a philosopher whose vision of democracy helped shape the political structure of the emerging nation, he was also a farmer, a scientist, and a revolutionary whose feet were firmly planted in the American ethos. So it isn't surprising that Jefferson would find uses for graphical displays that were considerably more down-to-earth than the life spans of heroes from classical antiquity. What is surprising is that he found time, while he was president of the United States, to keep a keen eye on the availability of thirty-seven varieties of vegetables in the Washington market and to compile a chart of his findings (a detail of which is shown in figure 6.2).

Henry Randall, Jefferson's nineteenth-century biographer, found that the "leader of a great civil nation—the founder of a new party and creed—the statesman engaged in the pressing affairs of a nation—watching with a green-grocer's assiduity, and recording with more than a green-grocer's precision, the first and last appearance of radishes, squashes, cabbages, and cauliflowers in the market—suggests a curious train of reflections!" What may have seemed curious to Randall seems perfectly aligned with the democratic American spirit that was so well represented in Jefferson.

theories of industry, business, and politics. His polemical political tracts argue pragmatically rather than moralistically. Another feature of Playfair's scientific orientation is discernible in the biography of his older brother, John, who took charge of William's education when the latter was thirteen years old. The older Playfair was a member in very good standing of the scholarly establishment, with prominent achievements in mathematics and geology.

The Graphical Inventions of Dubourg and Ferguson: Two Precursors to William Playfair

Eighteenth-century Europe was a hotbed of invention and discovery. It was the time of Bayes and Bernoulli, of DeMoivre and Euler, of Lagrange, Laplace, and Legendre. And although he did not begin publishing until the nineteenth century, the eighteenth century gets credit for the birth of Gauss. During this time, despite the absence of the kinds of modern communication we have all come to depend on, there seems to have been a tight network of communication among the intellectual elite of Europe. This network even expanded wide enough to include some of the brighter lights of the new world. This is well illustrated through the exploration of the development of one important graphical invention, the time line.

William Playfair gave prior credit to only one person in the invention of graphical displays of data: Joseph Priestly, for his use of the time line to show the lives of major figures from classical antiquity. Yet others, almost surely known to Playfair, preceded even Priestley. This is the story of two of those others.

In the preface to his 1786 *Atlas*, Playfair wrote,

> On inspecting any one of these Charts attentively, a sufficiently distinct impression will be made, to remain unimpaired for a considerable time, and the idea which does remain will be simple and complete, at once including the duration and amount. Men of great rank, or active business, can only pay attention to general outlines; nor is attention to particulars of use, any further than as they give a general information: And it is hoped, that with the assistance of these Charts, such information will be got, without the fatigue and trouble of studying the particulars of which it is composed.

Playfair certainly had a clear understanding of the power of graphical inventions, but he was not the first. The idea of graphical display was in the air of Western Europe at this time. In this chapter we will introduce two more precursors to Playfair. I have described Joseph Priestley's 1765 time line (chapter 5) as well as Thomas Jefferson's later use of the same

The content of much of this chapter is taken from the detailed and wonderful essay by Stephen Ferguson cited in the references. Figures 7.1 and 7.2 are photographs of the Dubourg scroll in the rare books collection of the Princeton University Library. Figure 7.3 is taken from the *Encyclopaedia Britannica*. I am grateful for the permission from all of these sources to reproduce what I have in this chapter. I am especially grateful to Stephen Ferguson for bringing Dubourg's work to my attention and for his scholarship in explicating it.

format (chapter 6). This chapter adds two more, quite marvelous examples: one by the French physician Jacques Barbeau-Dubourg (1709–1779) and the other by the Scottish philosopher Adam Ferguson (1723–1816).

In the third volume of Diderot's *Encyclopédie*, following "Chapeau" and "Choréographie," was an article by Diderot titled "Chronologique (machine)." The machine he was discussing was not a clock (or it would have been described later under "Horloge"), but rather a graph. More particularly, a graphic in the form of a fifty-four-foot scroll configured in a way not unlike a torah, which was produced by Dubourg in 1753. The "machine" that Diderot referred to was the housing for the scroll (see figure 7.1), which protected it when it was closed and served as a platform for easy viewing when opened. When the case was opened, the reader was faced with 150 years of history.

The scroll is a complex time line spanning the 6,480 years (one year equals 0.1 inch) from The Creation until Dubourg's time. This is demarked as a long thin line at the top of the scroll, with the years marked off vertically in small, equal, one-year increments. Below the time line, Dubourg has laid out his record of world history. He includes the names of kings, queens, assassins, sages, and many others, as well as short phrases summarizing events of consequence. They are fixed in their proper place in time horizontally and grouped vertically either by their country of origin or

Figure 7.1. Photo of Jacques Barbeau-Dubourg's scroll closed and latched. Princeton University Library, Department of Rare Books and Special Collections.

in Dubourg's catch-all category at the bottom of the chart "événements mémorables." An excerpt is shown in figure 7.2.

Why did Dubourg do this? In fact his motivation was quite similar to Playfair's, and expressed in comparable fashion. He declared that history has two ancillary fields: geography and chronology. Of the two, he believed that geography was the more developed as a means for studying history, calling it "lively, convenient, attractive." By comparison, he characterizes chronology as "dry, laborious, unprofitable, offering the spirit a welter of repulsive dates, a prodigious multitude of numbers which burden the memory." He believed that by wedding the methods of geography to the data of chronology he could make the latter as accessible as the former. Dubourg's name for his invention, *chronographie*, tells a great deal about what he intended, derived as it is from *chronos* (time) and *graphein* (to write). Dubourg intended to provide the means for chronology to be a science that, like geography, speaks to the eyes and the imagination, "a picture moving and animated." In this attitude he foreshadowed Bertin's distinction between reading and seeing by more than two hundred years.[1]

Dubourg's *chronographie* is marvelous not just for its form but also for its content. In

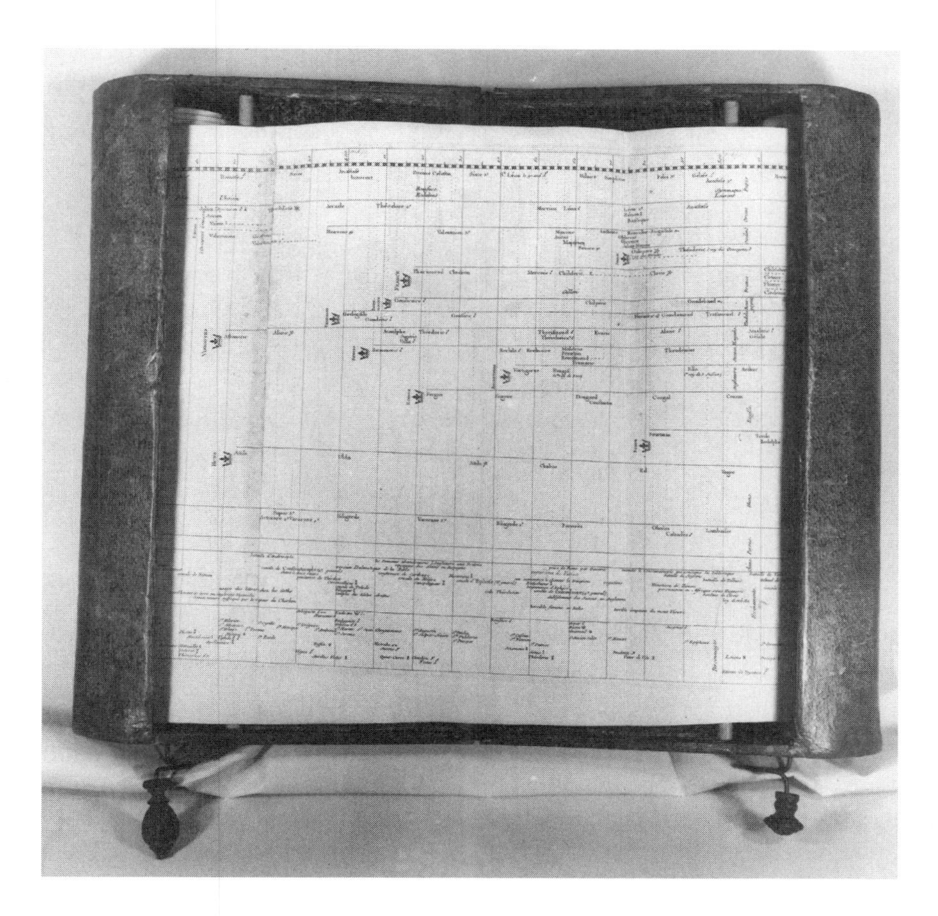

Figure 7.2. Photo of Dubourg's scroll opened to the years A.D. 360 to 510. Princeton University Library, Department of Rare Books and Special Collections.

addition to providing the names of various principal characters in the march of history, he includes a symbol that denotes character (martyr, usurper, tyrant, just, bigot, cruel, debaucher, slothful, fool, noble, majestic, blessed, heretic, impious, upright, unfortunate, rebel) as well as profession (savant, painter, theologian, botanist, physician, musician, monk, soldier, astronomer). Dubourg's annotations reveal his teaching agenda: he thought that the study of history was the path to virtue. By studying Dubourg's symbols, the student could answer the question, "what sort of person was King so-and-so?"

An interesting sidelight on the intellectual life of eighteenth-century Europe is provided in a letter dated May 8, 1768, from Dubourg to Benjamin Franklin (they had been friends for twenty-five years) about Dubourg's graphical invention. Apparently, Franklin had sent to Dubourg a copy of Priestley's biographical chart (see chapter 5), which shared the graphical character of Dubourg's but was published more than a decade later. Dubourg wrote, "I have the honor of sending to you the attached short explication of my chronographical chart which you have the goodness to ask of me. . . . I have received with gratitude and viewed with pleasure the biographical chart of Mr. Priestley which is in effect made according to almost the same principles as mine—without plagiarism on either part, because I do not intend to pride myself on the [earlier] date of mine."

As an historical aside, bearing again on the intertwining of eighteenth-century intellectuals, we note that Priestley dedicated the 1769 elaboration of his *New Chart of History* to Benjamin Franklin. Having already documented the connection between Dubourg and Thomas Jefferson (chapter 6), we complete the American connection with a note from John Adams's *Diary*, affirming that on May 20, 1778, he "dined this day at Dr. Dubourg's, with a small company," in which he goes on at length to describe the emotions he felt while looking at the painting on the ceiling of the gallery of Versailles.

Considerable evidence suggests that William Playfair got at least the germ of his graphical ideas from his older brother John. Moreover, both Playfairs knew of Priestley's time-line graphics. Thus it should not be surprising to discover that the first time line to find its way into an encyclopedia (the description of Dubourg's by Diderot does not count) was produced by someone connected with the Playfairs. Adam Ferguson (1723–1816), a Scottish historian and philosopher, came to Edinburgh to succeed David Hume as keeper of the Advocates' Library. He became the professor of natural philosophy at the University of Edinburgh in 1759.* Thus he overlapped with John Playfair at Edinburgh from 1770 until Ferguson's retirement in 1785. In 1780 he published a time line of the birth and death of civilizations that begins at the time of the Great Flood (2344 B.C.—indicating clearly, though, that this was 1,656 years after The Creation). This appeared in the second edition of the *Encyclopaedia Britannica* and is reproduced in figure 7.3.

Ferguson's time line shares some of the form and content of Dubourg's, but focuses on the character of entire civilizations rather than on individuals. The years run down the

* Adam Ferguson is a member of the Edinburgh group of "Scottish moralists" that included Hume, Lord Kames, and Adam Smith. He has been called the "first sociologist." Ferguson's chief significance is due to the prominent part he played in divorcing philosophy from the prevailing rationalistic, a priori approach and in substituting the inductive historical method. John Playfair played a similar role in geology.

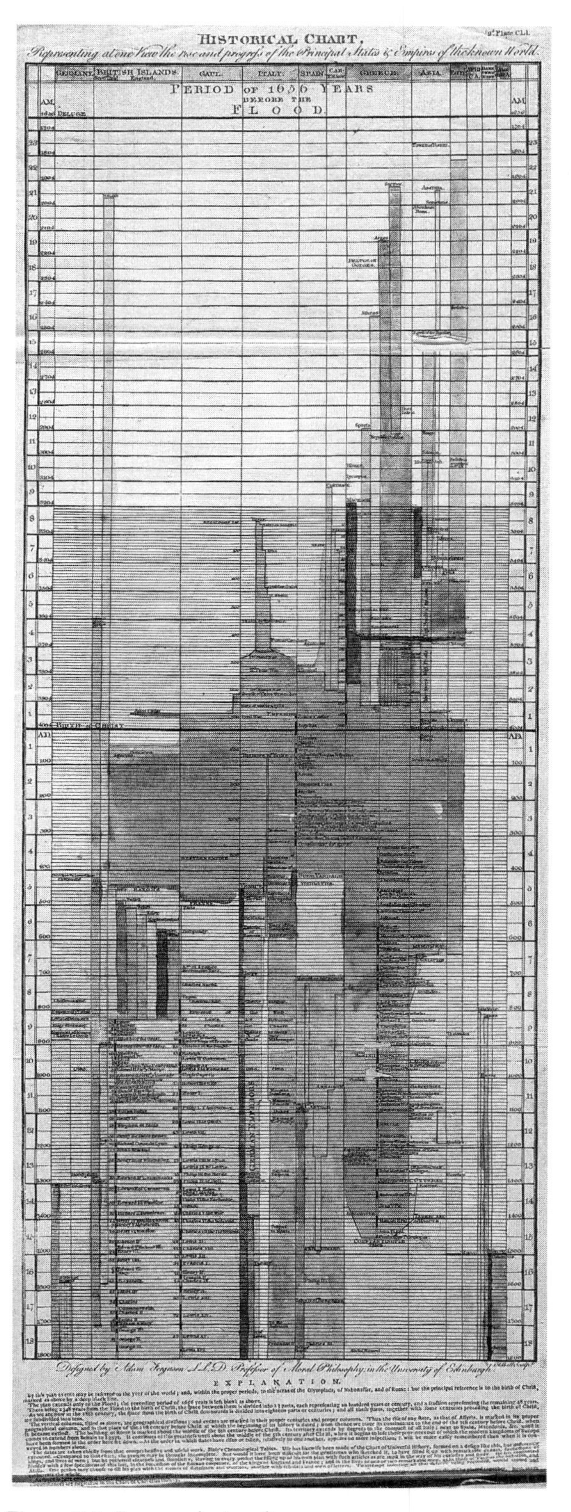

Figure 7.3. Ferguson's time line.

page, and the countries containing the civilization run across it. The order of the countries is critical because the spread of a civilization (e.g., the Roman Empire) is shown by expanding the colored shading associated with that civilization across borders. A careful examination of Ferguson's chart reveals the intelligence and care that went into its preparation.

The story surrounding the two time-charts described here points out how closely tied were the intellectual developments of Western Europe in the eighteenth century. The major figures all knew one another, and intellectuals from the American colonies were an integral part of the mix. It is no wonder that this aggregation of scholars thought of themselves as the "Republic of Letters": their fellowship ignored national bounds. In one sense it was the same as it is now, but instead of rapid electronic communication, our forebears made due with the post and by eating dinner together at every opportunity. But in another sense it was different. Modern intellectuals are closely connected with others in their fields, whereas those of the eighteenth century seemed to cross professional bounds with impunity. Perhaps there is yet something else we can learn from them about the fruits of cross-fertilization.

8 Winds across Europe: Francis Galton and the Graphic Discovery of Weather Patterns

The old adage about everyone talking about the weather but nobody doing anything about it needs to be generalized. Historically, people not only talked about the weather but also graphed it. Indeed, the development of graphic displays of data has been intertwined with the study of the weather for more than three hundred years. Christopher Wren (ca. 1663) developed an intricate weather clock that graphically recorded wind direction (in polar coordinates) and temperature (in rectangular coordinates) on a common time axis, as well as segregating hourly rainfall collections.* Twenty years later, Martin Lister presented various versions of graphical summaries of weather data before the Oxford Philosophical Society (March 10, 1683) and, later in the same year, a modified version to the Royal Society. In his audience sat Robert Plot, who was evidently enthralled with the possibilities of using Lister's method to provide a coherent rendering of complex information. He subsequently recorded the barometric pressure in Oxford every day of 1684 and summarized his findings in his "History of the Weather" graph (figure 1.5), discussed in chapter 1. The insight he shared with Lister in his 1685 letter (quoted in chapter 1) suggests explicitly that appropriate data, properly displayed, would provide answers to important questions.

I was reminded of this bit of graphical history during some e-mail correspondence with the eminent Chicago statistician (and historian) Stephen Stigler, who asked how often graphs ever really help us to make discoveries

> Graphs are wonderful for display but historically (I conjecture) they mostly portray what the investigator knows is there. They state, they do not (generally) discover. Why would that be? Because the world is still too highly dimensional to permit blind graphical searches with a high probability of success, a very high degree of guidance is needed. The mind + mathematics still tend to be superior as a discovery tool; graphs excel as rhetorical devices (using rhetoric in the best sense).
>
> I don't really mean to run down graphs (I always emphasize graphical diagnostics in teaching) but I still hold out for some vivid scientific discoveries where the graph came first. Many examples were known before they were picked up as vivid examples in recent times. I mean fancier graphs—simple marginal plots I do accept as extremely useful.

He then asked whether I could provide an example of a scientific discovery made graphically. First, I suggested John Snow's plot of

* A drawing of Wren's invention is at the Royal Institution of British Architects and is described in Hoff and Geddes (1962).

the September 1854 London cholera epidemic, which showed that the water from the Broad Street pump was the probable cause.* But he dismissed that example, saying, "that was a sort of guided search; he found more or less what he was looking for." I countered with the relatively recent discovery of the errors in the London *Bills of Mortality*.† Stigler tersely replied, "no, no."‡

But why not? I thought to myself, is the "discovery" too small? Or have we come to expect clerical errors? I am fond of this example because neither Graunt nor Arbuthnot found these mistakes (maybe not even Huygens), which would have been visible in a flash, had they known to draw a graph. I was reminded of a story that the Princeton polymath John Tukey once told me about an old mathematician who used to classify all theorems as either "impossible" or "trivial." Tukey once brought him a theorem that was quickly categorized as impossible. Tukey then proved it. The response immediately changed to, "Oh yes, it is trivial." The finding of the errors was trivial once a graph was drawn, but was totally missed by those worthy gentlemen.

As yet another example, I suggested the map of the Earth that shows how the east coasts of North and South America fit neatly into the west coasts of Europe and Africa, immediately suggesting plate tectonics. But this worthy suggestion also garnered the same "no, no."

It was at this point, perhaps sensing my growing frustration, that Stigler suggested Galton's discovery of the anticyclonic movement of the wind around a low-pressure zone (in the northern hemisphere). He said that it was "totally unexpected and purely the product of his remarkable high-dimensional graphs."

As it turned out, Galton's efforts to understand weather represented the beginning of the world that Robert Plot's prescient letter to Martin Lister imagined.§ Indeed, some might be moved to use this as an illustration of nineteenth-century discovery resulting from the concatenation of seventeenth-century genius with nineteenth-century industry.

In his 1863 book *Meteorographica*, Galton reports the results of a data gathering effort from a mixture of weather stations, lighthouses, and observatories across the face of Europe, spanning an area 1,500 miles east to west and 1,200 miles north to south. Distributed in this area were more than three hundred observers. Over the course of December 1861, Galton retrieved from them thrice-daily measurements of wind direction and velocity, barometric pressure, temperature, cloud conditions, and moisture. He adjusted these figures for altitude and constructed a multivariate glyph for each recording station. These glyphs were then implanted on a map of Europe. He made ninety-three

* This plot is well known and is considered the start of modern epidemiology (see Wainer 2000b, p. 61). Dr. Snow and his famous graph will join us in chapter 21 (figure 21.1).

† The London *Bills of Mortality*, discussed at some length in the introduction to this book as well as in chapter 1, were a sixteenth-century response to the bubonic plague. The plague, an infectious disease transmitted by fleas carried by rodents, recurred frequently for nearly four hundred years. The vectors of contagion were unknown at the time, and it was generally believed that the contagion was airborne. Contagion was reduced through quarantine. In the 1530s a warning system was instituted in London by requiring parish priests to submit weekly reports on the number of plague deaths. This system became continuous on the December 29, 1603, and the individual reports were collected into a book and analyzed by John Graunt in 1662. The Chicago statistician Sandy Zabell (1976) made graphs of the data contained in Graunt's book and discovered clerical errors and outliers that had lain dormant for more than three hundred years.

‡ Stigler's view is not universally held among statisticians. Harvard's distinguished professor Fred Mosteller wrote to me (April 16, 2002), "I would have said 'Yes-Yes' to London Bills of Mortality" and to the fitting together of coastlines.

§ Galton's work was clearly the intellectual offspring of Lister's. See chapter 1 for more on Lister and his followers.

such maps, three per day (morning, afternoon, and evening) for each of the thirty-one days of December. These maps formed the basis of his study of the weather. A sample map, from the afternoon of December 3, 1861, is shown as figure 8.1. The U-shaped icons show the direction of the wind, and the extent to which they are filled in shows its velocity (more fill indicates greater velocity). On the facing page, Galton produced a map on which were plotted icons showing the barometric pressure (see figure 8.2). The perceptual joining together of these two kinds of figures provided the grist for his discoveries.

Galton's study of these ninety-three annotated maps began to yield the sorts of global understanding of the weather that was predicted by Plot's prescient view quoted in chapter 1 (p. 15). Galton summarized his findings by first pointing out that, as large as the geographic area of his study was, weather systems are much larger still: "Bearing in mind the vastness of [the area studied] it may well astonish us to find frequent wind currents

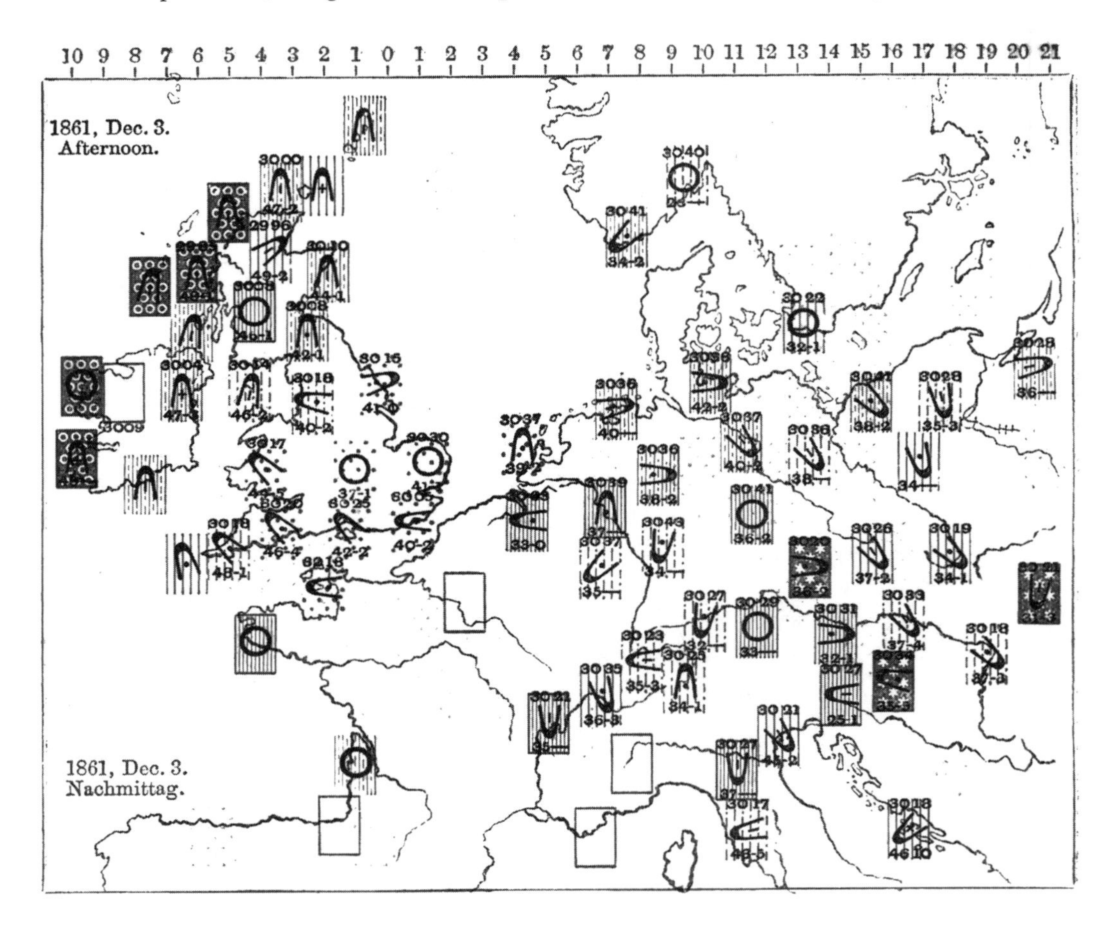

Figure 8.1. Reproduction of Galton's (1863b) graph of the wind patterns over Western Europe on the afternoon of December 3, 1861. The U-shaped symbol shows the direction of wind; the amount of fill indicates the strength. An O-shaped symbol indicates calm. One can see the clockwise movement of the wind around the high pressure centered in Germany (see figure 8.2). Princeton University Library, Department of Rare Books and Special Collections.

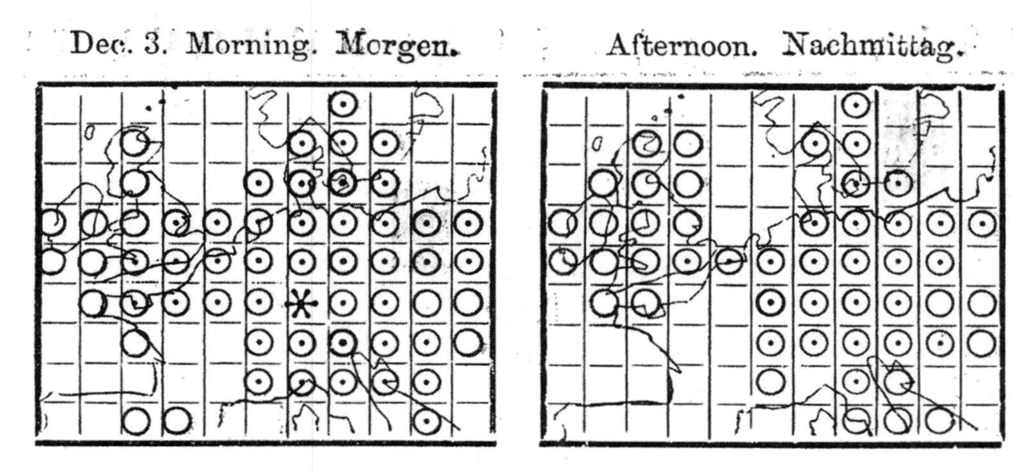

Figure 8.2. Reproduction of the barometric readings on the morning and afternoon of December 3, 1861. The circles with a dot in them represent readings of 30.21 to 30.45 inches of mercury; with no dot, 29.96 to 30.20; the star, 30.46 to 30.70 (from Galton 1863b).

sweeping with an unbroken flow across our Charts, being obviously portions of enormous systems which far overpass their limits, and probably connect the winds of the tropics on the one hand with those of the Arctic circle on the other." Using the longitudinal nature of the sequential charts, he also examined the speed with which weather changes: "Still more surprising is it to remark the simultaneity of the wind changes. We may see systems of currents converting themselves into entirely different combinations, with perfect unanimity throughout the whole area over which our observations extend, . . . testifying to the remarkable mobility of the air. It is difficult not to conclude, even from these limited series of weather maps, that ordinary changes of wind and sky have their sources in far more numerous and distant regions than is commonly supposed." And, after noting that "the Alps form a barrier which the winds rarely overleap without change of direction," he concluded that "these maps bear testimony to the shallowness of that lowest stratum of the atmosphere, the movement of which causes our winds."

Galton mentioned that many meteorologists were eager to see these charts in the hope of confirming or opposing the "theory of cyclones." He then went on to point out that they testify not only to the existence of cyclones, but also to what he called "anticyclones." This is the counterclockwise movement of air around a low-pressure zone. After seeing evidence for the existence of anticyclonic movement, Galton concluded that their existence "is consonant with what we should expect. It is hardly possible to conceive masses of air rotating in a retrograde sense in close proximity, as cyclonogists suppose, without an intermediate area of direct rotation, which would, to use a mechanical simile, be in gear with both of them, and make the movements of the entire system correlative and harmonious."[1] He summarized this in the schematic representation shown in figure 8.3.

Thus, not only did it take almost two hundred years for Plot's vision of the marriage of graphical methods and meteorological data to occur, it also took Francis Galton to do it. And although we are still unable to predict "plague, and other epidemical distempers," Galton's

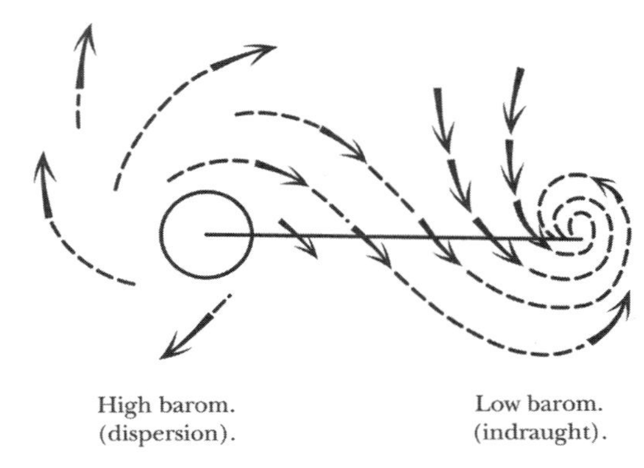

Figure 8.3. Galton's graphical conclusion showing the coordinated pathways of wind around a high pressure–low pressure system derived from looking at ninety-three pairs of such graphs as shown in figures 8.1 and 8.2. This was Galton's (1863a) figure 2.

exercise demonstrated how the combination of well-chosen data with good display can yield rich and unexpected understanding.

Seeing what Galton was able to accomplish from one month's data cleverly summarized with multivariate graphic icons dispersed over a map brings to mind what R. A. Fisher accomplished sixty years later. In 1919, Sir John Russell appointed Fisher as the statistician at the Rothamstead agricultural station, where Fisher inherited ninety years of data from "experiments" that tested various combinations of mineral salts and various strains of wheat, rye, barley, and potatoes. These data were daily records of temperature and rainfall, information about fertilizer dressings and soil, and annual recordings of harvests. Fisher referred to his job as "raking over the muck heap." Indeed, since there was little in the way of a coherent idea of experimental design in the generation and collection of these data, "muck heap" seemed an accurate characterization. Yet Fisher, like Galton, was able to draw inferences that no one before him could. He did this by organizing the data and inventing suitable methods of analysis along the way. In addition to seeing what he could discover, he was also able to pinpoint what was unknowable from data in this form. This led him to lay out designs for future work that would permit these inferences. The parallels to Galton and the data collection for modern weather forecasting are obvious.

But where are we with respect to Stigler's question? Fisher's accomplishments seem clearly in the mode of "mind + mathematics," but what about Galton's? Obviously, his work might be best characterized as "mind + graphics," with perhaps a little mathematics thrown in as well. But were his discoveries entirely unexpected? Alas, I fear not. Indeed, Galton himself refers to "Dove's law of gyration," which on theoretical grounds predicts the anticyclonic movement. He then goes on to explain his month-long experiment, and reports that "I found it more or less present on from fifty to sixty occasions. Its existence is consonant with what we should expect."[2]

Is "entirely unexpected" too stringent a requirement? Is any discovery ever entirely unexpected? John Tukey has taught us that graphical displays are the best way to find what we were not expecting, but surely such discoveries usually take place in the region of observations about which we have some, perhaps competing, hypotheses. Galton's discoveries were unexpected to some extent, but I believe that his real contribution was to demonstrate the value of visually organized data collection for discovery, hypothesis generation, and hypothesis confirmation. By adjoining data to theory through his graphical inventions, Galton did for the collectors of weather data what Kepler did for Tycho Brahe. This is no small accomplishment.

II Using Graphical Displays to Understand the Modern World

The twenty-first century is overrun with data, often complex and more often voluminous.* The study of statistics, because it is crucial in the understanding of the modern world, has become much more common than in the past and is now taught widely in high schools (see figure II.1). As we have seen, graphic display fits snuggly with statistical data to allow us to see what we were not expecting. As with most things graphical, Playfair understood that viewers could understand a complex phenomenon presented graphically that might elude them if presented any other way. Hence when he tried to point out the profound effect the exponential growth that is the soul of compound interest has on England's national debt, he chose to communicate it in a graph (see figure II.2). Moreover, he also understood how hard it would be to achieve the same apparently affect-free impact of a picture if he had to use just words.

In this part I describe some discoveries that are easily made graphically but can remain effectively hidden otherwise. The stories are varied in topic but are all generally short. They are frequently worthy of deeper investigation but in only the first story (chapters 9 and 10) does my resistance to greater depth collapse. Chapter 9 contains the remarkable result that during the Vietnam War, the average test scores for both those who entered the military and for those who did not were lower than before the war. This is an example of a common phenomenon that is known today as Simpson's Paradox.† Statistical paradoxes are common and, alas, fiendishly easy to fall prey to. Thus I felt justified in deviating from the bright path of graphical display and spending chapter 10 illustrating and discussing two common paradoxes.

The balance of the part comprises a collection of graphical gotchas demonstrating how some simple graphical procedures help us to understand the world we live in. The examples range from uncovering the character of the U.S. Supreme Court (chapter 11) to understanding the ethnic character of a jury pool (chapter 12). I also use graphs to

* Despite what we hear in political speeches and see in television exposés, the plural of *anecdote* is not *data*. Hence we might amend this sentence to be more complete by adding "and nondata." Richard Saul Wurman (1976) expressed this in his oft-quoted aphorism, "Everyone spoke of an information overload, but what there was in fact was a non-information overload." In this part we focus on data display, but we will stray into a more detailed discussion of nondata in chapter 22.

† This aggregation paradox is named after Edward Hugh Simpson, who described it in a 1951 paper (Simpson 1951), although it was recognized by the Scottish statistician George Udny Yule almost fifty years earlier. The finding of a phenomenon that is named after the wrong person is quite common and is known as Stigler's Law of Eponymy, which, simply stated, is that every invention or discovery is inevitably named after the wrong person. Stigler (1980) credits this "law" to Robert Merton.

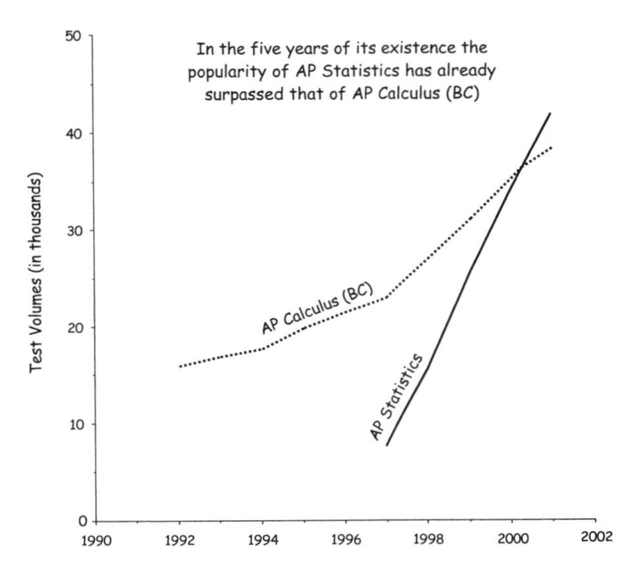

Figure II.1. The spectacular growth in the popularity of the study of statistics reflects the growing realization of its importance in the modern world.

study shipping behavior (chapter 13), trends in the stock market (chapter 14), sex, smoking, life insurance, and convertibles (chapter 15), and some graphic malfeasance in the *New York Times* (chapter 16). In chapter 17, I show how a graphical analysis of sex differences in athletic performance clarifies one's intuition, and in chapter 18, I emphasize how focusing attention on the essentials of what the viewer of a display wants to know provides guidance toward its construction. This is illustrated with a standard college entrance exams score report sent annually to more than a million high school students and a suggested modification.

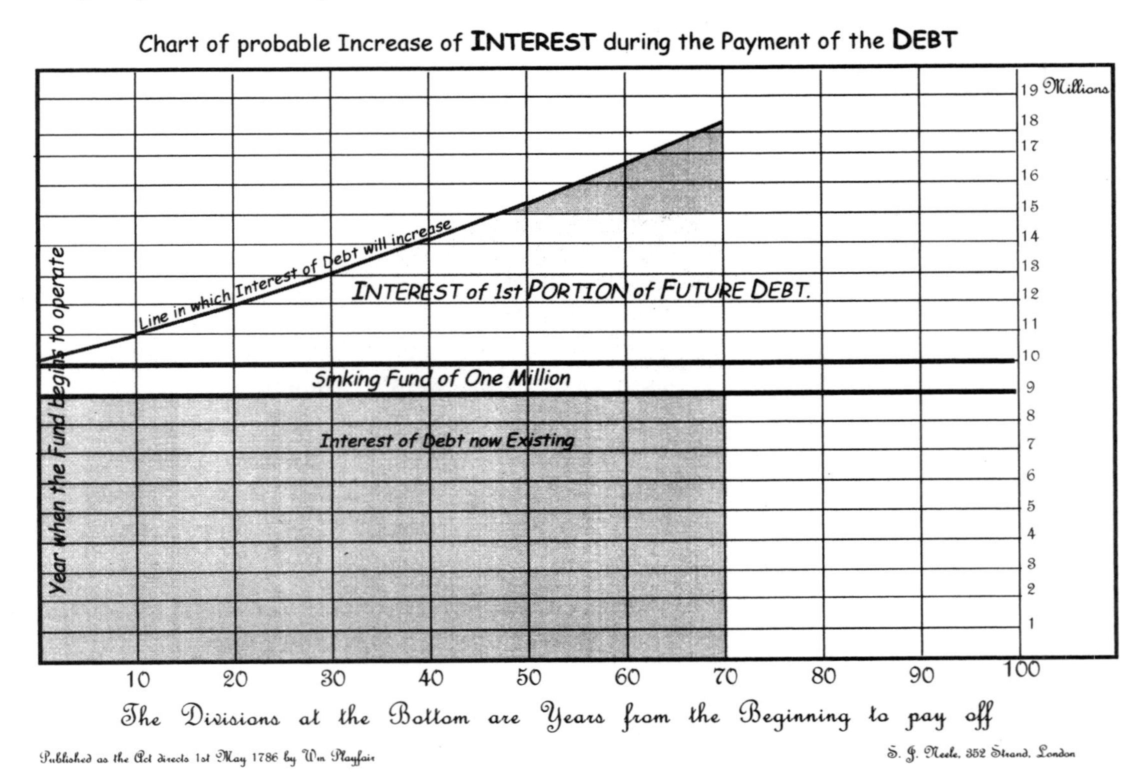

Figure II.2. William Playfair's use of a graphical representation of the effect of compound interest indicates that he understood the power of his method to effectively communicate complex quantitative phenomena.

9 A Graphical Investigation of the Scourge of Vietnam

When I moved from Princeton to Philadelphia, I raised the average income in both places. Those unfamiliar with statistical artifact may have to ponder momentarily how such an event might come to pass. But once one has done so, the implication is clear. In this chapter, I point out a similar result—specifically, how the Vietnam War managed to lower the average ability both inside and outside the American military.

Yogi Berra pointed out, "You can see a lot just by looking." I'm sure that Yogi Berra had other things on his mind when he uttered his now famous line, but that does not change its broad applicability. A graphic display has many purposes, but it achieves its highest value when it forces us to see what we were not expecting. This was brought home to me again recently when I was looking at a table of data showing the number of SAT exams that had been administered annually by the Educational Testing Service (ETS). Being of a visual bent, I immediately drew a graph of these data (see figure 9.1) and saw first that the test volumes had turned upward recently after having been pretty steady over the last decade or two. But a giant lump in the display could not help but catch my eye.

It was easy to see that the 60 percent increase in SAT volumes coincided with the Vietnam War. Thus, one explanation for this increase is that large numbers of young men sought student deferments to avoid military service. This interpretation gains in credibility when, in 1970, student deferments were eliminated and SAT volumes dropped just as precipitously as they had risen. Because these data began in 1960, I was uncertain whether this was a Vietnam effect or a more general "war effect." This question sent me to the ETS archives, where I was able to retrieve test volumes for the SAT (or for the college board tests that preceded it) since its origin in 1901. In its first year, the College Entrance Examination Board administered essay exams in nine subjects to 973 candidates at 69 testing centers. By 1969, the number of exams administered exceeded a million and a half at more than 4,500 testing centers. To show this entire data series with reasonable resolution, it seems sensible to use a logarithmic scale. This is shown in figure 9.2.

The Vietnam lump is still visible, although it appears less dramatic on the log scale. We also see an enormous decline in test volumes in the fifteen-year period that spans the

Depression and World War II. Several things are apparent:

- The economic hard times of the early 1930s had a serious effect on college enrollments,
- World War II was different than Vietnam in terms of the willingness of young men to participate, and
- The end of World War II, with the aid of the GI Bill, marked the beginning of a boom for college enrollments that continued unabated for twenty-five years.

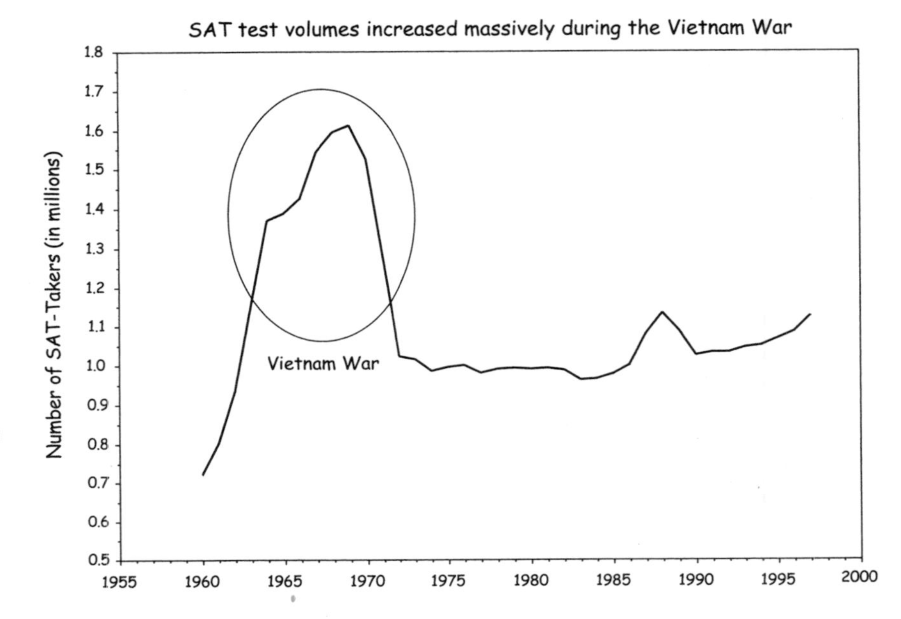

Figure 9.1. The annual test volumes for the SAT since 1960. The huge increase in SAT-takers during the Vietnam War is an obvious effect. Data source: *Annual Reports of the College Board.*

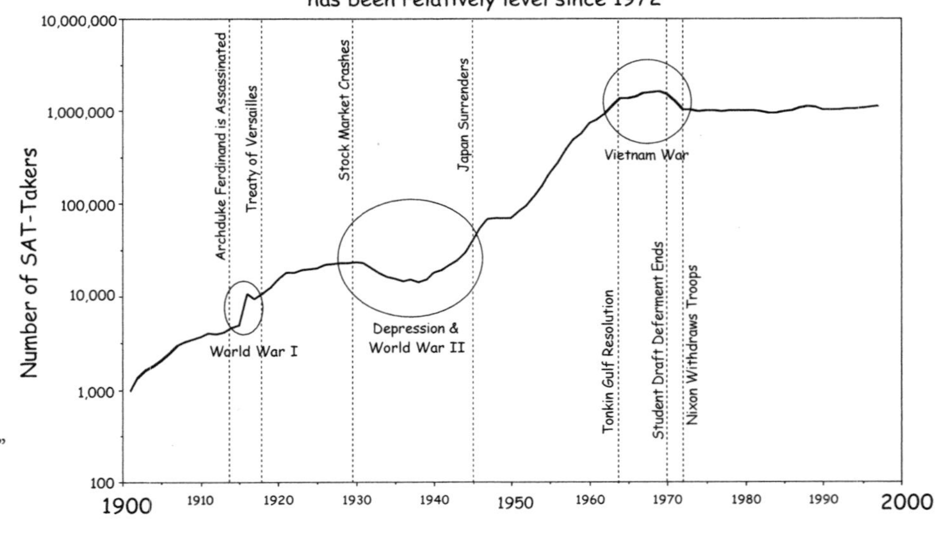

Figure 9.2. The annual test volumes for the SAT since its beginning in 1901. Data source: *Annual Reports of the College Board* and "SAT Candidate Volumes, 1926–1961," a memo prepared by Arthur Bogan of ETS.

Last, we can see a steep one-year increase (from 5,000 to 10,000) between 1915 and 1916, but it is unclear what might have caused it. What is clear from these data is that the Vietnam effect seems unusual. Might other data shed light on this phenomenon?

This question sent me scurrying for my trusty set of the Bureau of the Census's *Historical Statistics* books, where I found a data series describing the number of PhDs awarded annually since 1870. This series shows the same pattern (see figure 9.3), perhaps with even more emphasis. We see the same decline during World War II (for both

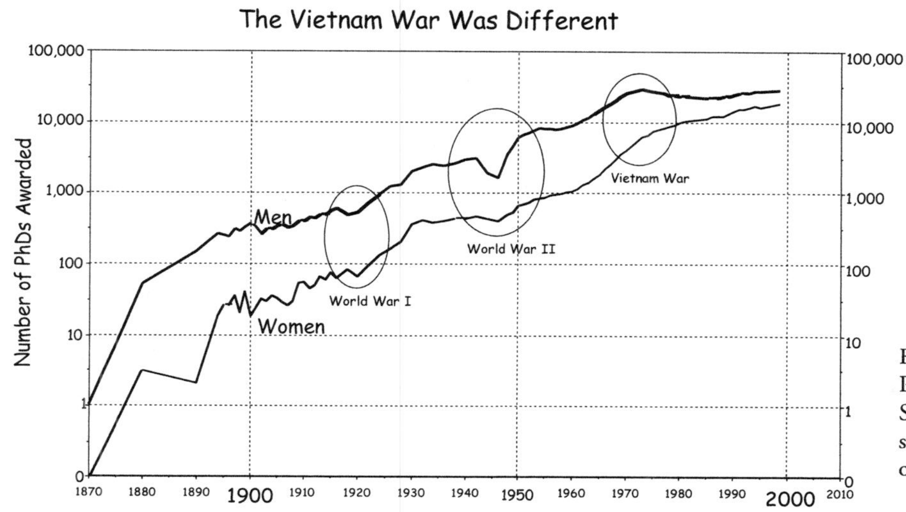

Figure 9.3. The number of PhDs awarded in the United States since 1870, shown by sex. Data source: U.S. Bureau of the Census (1989).

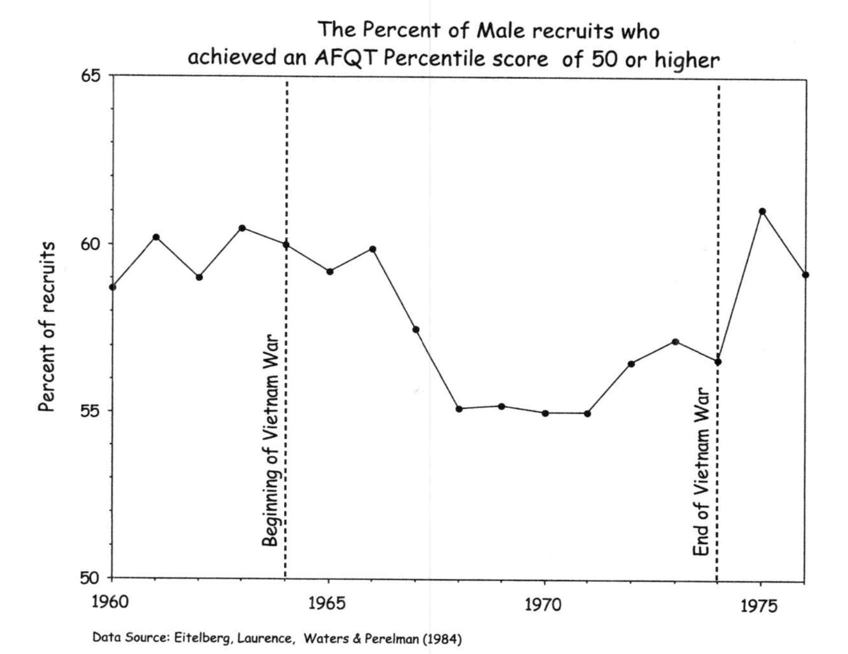

Figure 9.4. The percentage of U.S. military accessions who scored above the national median on the Armed Forces Qualifying Test (AFQT) from 1960 until 1976. Data source: Eitelberg, Laurence, Waters, and Perelman (1984).

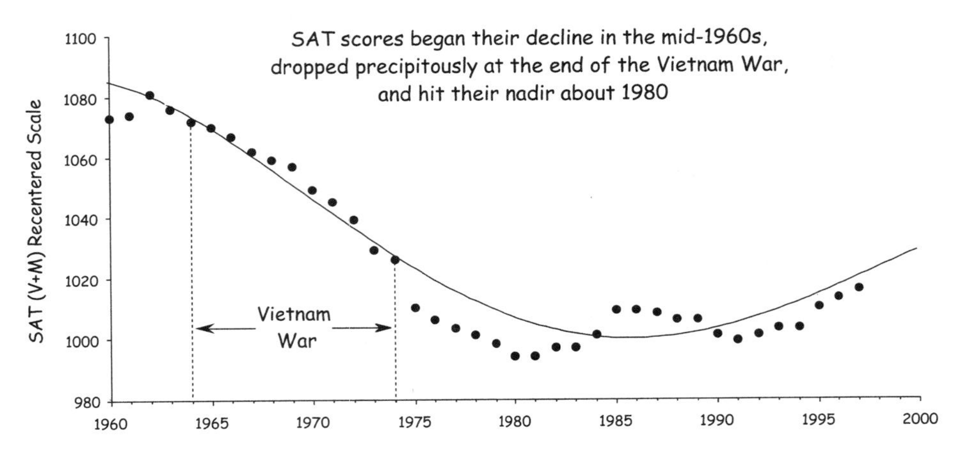

Figure 9.5. The average SAT score (verbal + math) among U.S. examinees from 1960 until 1997. Data source: College Board, *National College-Bound Seniors*, annual reports for various years.

men and women, although less pronounced for women). There is a matching decline during World War I, and an increase in PhD production during the Vietnam War (mine among them). Of course, these blips on the time series are minor effects, the major ones being the general growth patterns of PhDs and the shrinkage of the difference between men and women. Nevertheless, these blips support the phenomenon observed with SAT volumes.

But if there was an increase in college applicants (and hence almost surely college enrollments) during the Vietnam War, what did that do to the average ability within the military? All enlisted personnel are tested upon entry and their scores are compared with those obtained from a representative national sample. The results of that testing program are shown in figure 9.4. The dip in enlistees' ability during the Vietnam period is obvious.

So it appears that at least some of the cream of the military crop was siphoned off into the college population during the Vietnam War. How did that influx of fresh talent manifest itself within the college population? We see (in figure 9.5) that the principal decline in SAT scores over the last thirty-five years occurred during the Vietnam War.* An obvious explanation for this phenomenon brings to mind my initial quip. This seems to be ample evidence in support of our original hypothesis that the shift in college attendance associated with the Vietnam War managed to lower the average ability of both those who went into the military as well as those who did not. This effect seems specific to the Vietnam War; similar data suggest the opposite effect during World War II.

This seemingly anomalous situation—that scores went down in both groups—could occur even if the average score of everyone, taken together, went up! The details of this odd situation are the subject of the next chapter.

* An alternative explanation involving school prayer was proposed in chapter 4. I find an explanation involving participation rate more compelling.

10 Two Mind-Bending Statistical Paradoxes

Simpson's Paradox

In 1998, Ian Westbrooke described how New Zealand indigenous Maoris appeared to be overrepresented on juries in New Zealand: 9.5 percent of people living within the jury districts were Maori, compared with 10.1 percent of Maori in the pool of potential jurors.[1] Yet, upon looking more closely, Westbrooke found that "in every single local area Maori were underrepresented—often substantially." Such anomalies are not restricted to New Zealand.

Recently I have noticed a number of empirical results reported in the newspaper, like the one above, that seem, at least on the surface, to be self-contradictory. When I ask friends about the apparent anomalies, they snort and suggest that it must be a typo of some sort. As Westbrooke pointed out, these anomalies are examples of a common statistical artifact called Simpson's Paradox, which is when aggregated results point in the opposite direction as the same data when disaggregated.

We all view the world through the filter of our own experience. I am a statistician by training and vocation, and an empiricist down to my bones. I have trouble with arguments that do not contain empirical evidence, and those that do I look at with care. The newspaper is not the same for me as it is for most other readers.

On September 2, 1998, the *New York Times* reported evidence of high school grade inflation. They showed that a greater proportion of high school students were getting top grades at the same time that their SAT Math scores had declined (see table 10.1). Indeed, when we look at their table, the data seem to support this claim; at every grade level, SAT scores seem to have declined by 2 to 4 points over the decade of interest. Yet the article also reported that over the same time period (1988–1998), SAT Math scores had gone up by ten points.

How can everyone's scores go down while the average goes up? The key is the change in grade distribution of the children. For example, the proportion of children being awarded A+ grades in 1998 is almost twice that of a

Table 10.1
A Decade of High School Grades and SAT Scores: Are Students Getting Better or Worse?

Grade average	Percentage of students getting grades 1988	1998	Average SAT Math scores 1988	1998	Change
A+	4	7	632	629	-3
A	11	15	586	582	-4
A-	13	16	556	554	-2
B	53	48	490	487	-3
C	19	14	431	428	-3
		Overall average	**504**	**514**	**10**

Source: *New York Times*, September 2, 1998.

decade earlier. Thus, although it is true that SAT Math scores declined from 632 to 629 for A+ students, some of those A+ students might have been included in the A category if they had taken the test in 1988. When we calculate the average score, we weight the 629 by 7 percent in 1998 rather than by only 4 percent. The calculation of the average SAT score in a year needs to use both high school grades and SAT scores for students with those grades. One way for us to make the anomaly disappear is to artificially hold the proportional mix fixed. These data are indicative of likely grade inflation.

This anomaly is not rare. For example, consider the results from the National Assessment of Educational Progress shown in table 10.2. We see that eighth-grade students in Nebraska scored 6 points higher in mathematics than their counterparts in New Jersey. Yet we also see that white students do better in New Jersey. Black students do better in New Jersey, also. Indeed, all other students do better in New Jersey. How is this possible? Once again, it is an example of Simpson's Paradox. Because a much greater proportion of Nebraska's eighth-grade students (87 percent) are from the higher-scoring white population than in New Jersey (66 percent), their scores contribute more to the total.

Table 10.2
National Assessment of Educational Progress Eighth-Grade Math Scores (1992)

	Unadjusted state score	White	Black	Other nonwhite	Standardized state score
Nebraska	**277**	281	236	259	**271**
New Jersey	**271**	283	242	260	**273**
		Proportion of population			
Nebraska		87%	5%	8%	
New Jersey		66%	15%	19%	
Nation		69%	16%	15%	

Source: Mullis, Dossey, Owen, and Phillips (1993).

Is ranking states on such an overall score sensible? It depends on the question that these scores are being used to answer. If the question is something like "I want to open a business. In which state will I find a higher proportion of high-scoring math students to hire?" this unadjusted score is sensible. If, however, the question of interest is "I want to enroll my children in school. In which state are they likely to do better in math?" a different answer is required. If your children have a race (it doesn't matter what race), they are likely to do better in New Jersey. If questions of this latter type are the ones that are asked more frequently, it makes sense to adjust the total to reflect the correct answer. One way to do this is through the method of standardization, in which we calculate what each state's score would be based upon a common demographic mixture. In this instance, a sensible mixture to use is that of the nation overall. Thus, after standardization the result obtained is the score we would expect each state to have if it had the same demographic mix as the nation. When this is done, we find that New Jersey's score is not affected much (273 instead of 271), whereas Nebraska's score shrinks substantially (271 instead of 277).

Thomas D. Woolsey, whose data are shown in table 10.3, provided a third example of Simpson's Paradox. They show that, although the death rate in Maine (1,391) is higher than the death rate for the same year in South Carolina (1,289), the death rates for all age groups individually (except five- to nine-year-olds) are much higher in South Carolina. The answer, once again, is that Maine has a much higher proportion of older people, for whom the death rates are the highest. The extent of the distortion caused by the mismatched age distributions is seen when the death rates are

Table 10.3
Death Rates in Maine and South Carolina: Which State Is Safer?

	Percent of population in age category		Death rate (per 100,000)		
Age	Maine	South Carolina	Maine	South Carolina	Difference
0–4	9.4	11.8	2,056	2,392	-336
5–9	10.0	13.9	186	185	1
10–14	9.3	12.8	140	184	-44
15–19	8.6	12.2	223	426	-203
20–24	7.6	9.6	370	645	-275
25–34	13.3	12.6	391	871	-480
35–44	12.7	11.0	545	1,242	-697
45–54	11.3	8.3	1,085	1,994	-909
55–64	9.1	4.6	2,036	3,313	-1,277
65–74	5.8	2.3	5,219	6,147	-928
75+	2.8	1.0	13,645	14,136	-491
		Total death rate	**1,391**	**1,289**	**102**
		Adjusted death rate	**1,203**	**1,716**	**-513**

Source: Data from Woolsey (1947).

adjusted to reflect a common age distribution. That is, the death rate for each state, at each age grouping, is weighted by the same percentage (representation within the entire U.S. national population). At almost any age, it is a lot safer in Maine.

Stuart Baker of the National Cancer Institute provided my fourth and last example. He reports the results of an observational study in which one treatment was better for men and for women but inferior when aggregated over the two groups. A reduced version of his table 1 is shown in table 10.4. We see, looking across the first row, that 200 of the 240 men in the study were given treatment A, and that 120, or 60 percent of them, survived. In contrast, only 20 of the 40 (50 percent) who received treatment B survived. The second row shows the results for women. Although the overall survival rates for women were much higher, here too treatment A was superior (95 percent versus 85 percent). Yet, when the results for men and women are aggregated, we find that 72 percent of those who received treatment A survived, compared to 80 percent of those who received treatment B.

Simpson's Paradox is subtle, but in the more than two hundred years since Playfair drew his plot of the growth of compound interest, our experience has taught us that a graphic depiction often aids in our understanding of subtle phenomena. A graphic representation of Simpson's Paradox was provided by Baker and Kramer, who independently rediscovered a plot that had appeared in a Korean statistical journal fourteen years earlier.[2] Consider the graphic representation of the results from table 10.4 shown in figure 10.1.

The solid diagonal line shows the survival rate for those undergoing Treatment A for various proportions of women. At the extreme left, if no women were treated, the survival rate would be that for men, or 60 percent. At the extreme right is the survival rate if only

Table 10.4
Comparison of Survival Rates for Two Treatments

		Number getting treatment		Number surviving		Percentage surviving	
	Total	A	B	A	B	A	B
Men	240	200	40	120	20	60	50
Women	360	100	260	95	221	95	85
Combined	600	300	300	215	241	72	80

Source: Data from Baker and Kramer (2001), table 1.
Note: Percentage of those getting treatment A who are women is 33 percent [100 / (200 + 100)].
Percentage of those getting treatment B who are women is 87 percent [260 / (40 + 260)].

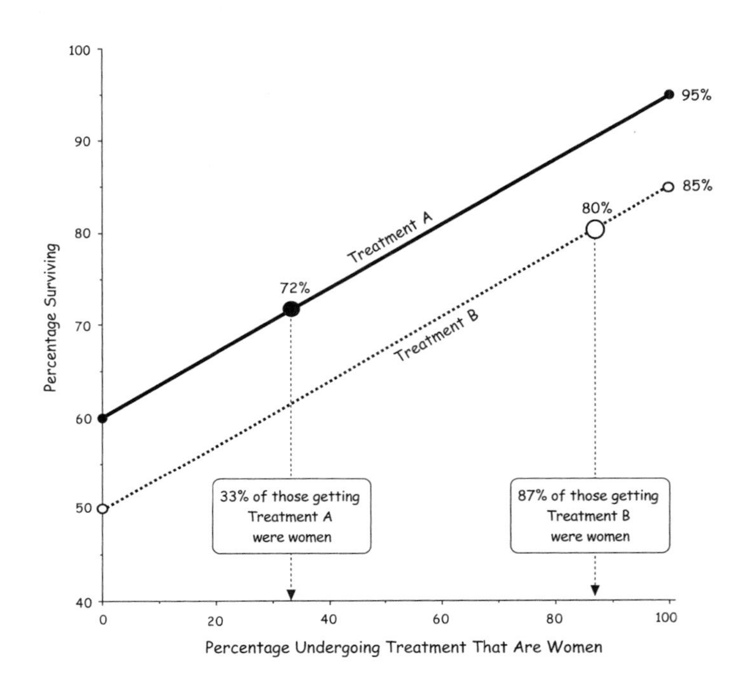

Figure 10.1. A graphical depiction of Simpson's Paradox. It shows that even though Treatment A is superior to Treatment B, the aggregation of differing proportions of men and women can yield the appearance of a reversal.

women were treated, 95 percent. The large black dot labeled "72%" represents the observed survival rate for the mixture that includes 33 percent women.

The dashed line shows the same things for Treatment B; the large open dot labeled "80%" denotes the observed situation in which 87 percent of those treated were women.

We see that for any fixed percentage of women on the horizontal axis, the advantage of Treatment A over Treatment B is the same, 10 percentage points. But because Treatment B garners a much larger proportion of more hardy females, its survival rate is higher than Treatment A's.

Now that we have this new tool, let us return to the Nebraska–New Jersey math competition discussed in the second example (figure 10.2). If we draw a solid line for Nebraska and a dashed one for New Jersey, we immediately see why the aggregated Nebraska score is higher. The small rectangle represents the location of the standardized racial mix (69 percent white) and the graph makes clear how and why standardization works—it uses the same location on the horizontal axis for all groups being compared.

Simpson's Paradox can occur whenever data are aggregated. If data are collapsed across a subclassification (such as grades, race, or age), the overall difference observed may not represent what is going on. Standardization can help correct this, but nothing short of random assignment of individuals to groups will prevent the possibility of yet another subclassification, as yet unidentified, changing things around again. But I believe that knowing of the possibility helps us, so that we can contain the enthusiasm of our impulsive first inferences.

Simpson's Paradox is well known among statisticians, but it is almost completely unknown to everyone else. I decided to include these examples in the hope that having them gathered in a single convenient place would help spread the word. In addition, the

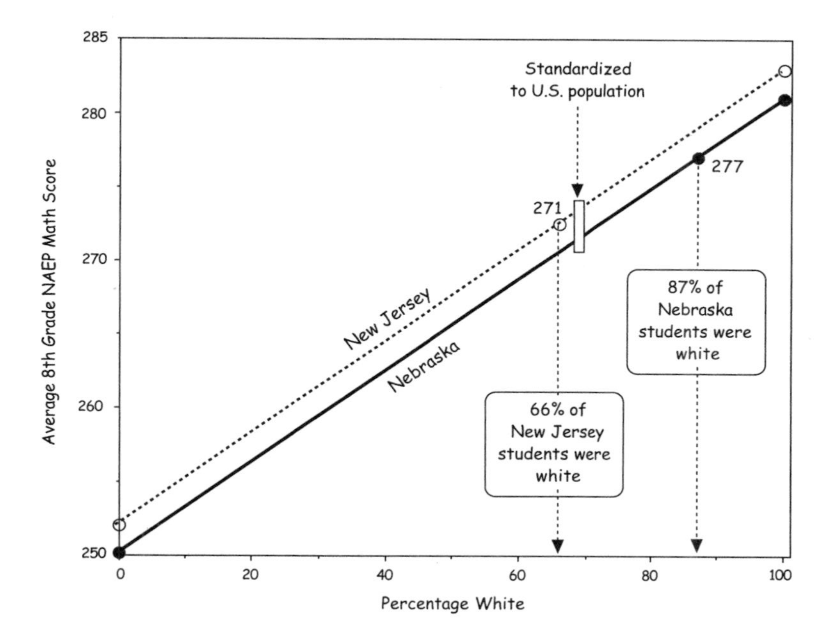

Figure 10.2. Using the Baker-Kramer plot to reexamine the New Jersey–Nebraska math data. It shows how, even although New Jersey's students outperform Nebraska's students, the aggregation of differing proportions yields the appearance of a reversal.

Baker-Kramer plot provides an easy summary of both the problem and its solution.

Kelley's Paradox

When we try to predict one event from another, we always find that the variation in the predictions is smaller than that found in the predictor. In 1889, Francis Galton pointed out that this always occurred whenever measurements were taken with imperfect precision. He called it "regression toward the average."* This effect is seen in some historical father-child IQ data shown in figure 10.3. The fathers' IQs vary over a 30-point range but their children's predicted IQs vary over only a 17-point range.

Although regression has been well understood by mathematical statisticians for more than a century, the terminology among appliers of statistical methods suggests that they thought of it either as a description of a statistical method or as only applying to biological processes. In 1924, the economic statistician Frederick C. Mills wrote, "the original meaning has no significance in most of its applications."[3]

Stephen Stigler pointed out that this was "a trap waiting for the unwary, who were legion."[4] The trap has been sprung many times. One spectacular instance of a statistician getting caught was "in 1933, when a Northwestern University professor named Horace Secrist unwittingly wrote a whole book on the subject, *The triumph of mediocrity in business*. In over 200 charts and tables, Secrist 'demonstrated' what he took to be an important economic phenomenon, one that likely lay at the root of the Great Depression: a tendency for firms to grow more mediocre over time."[5] Secrist showed that the firms with the highest earnings a decade earlier were currently performing only a little better than average; moreover, a collection of the more poorly

* More often he called it "regression toward mediocrity"; however, I choose to use the less colorful but more accurate term *average* because it is more descriptive.

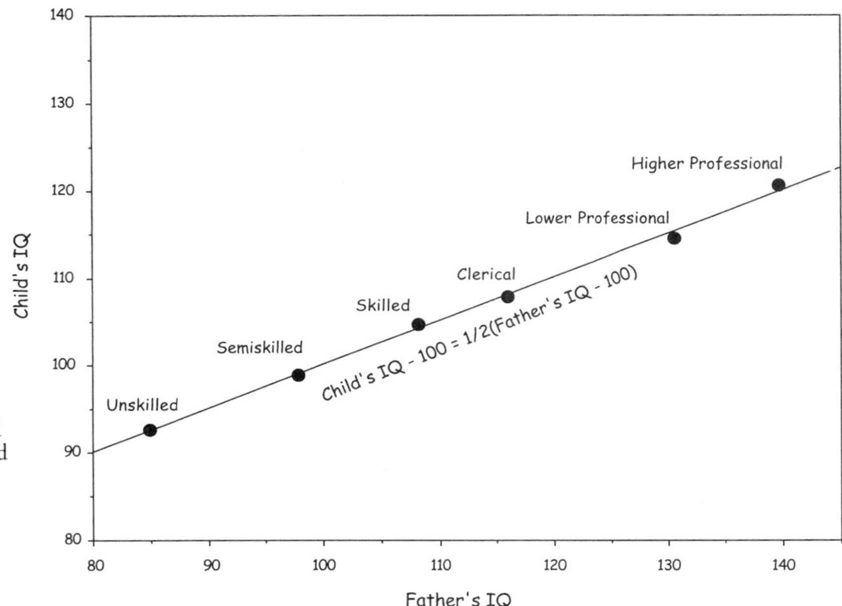

Figure 10.3. IQ scores of a child are roughly predictable from the IQ score of a parent. Yet the lowest- and highest-scoring adults are typically more extreme than most extreme-scoring children.

performing firms had improved to only slightly below average. These results formed the evidence supporting the title of the book.[6] The legendary Chicago statistician Harold Hotelling, in a devastating review published the same year, pointed out that the seeming convergence Secrist obtained was a "statistical fallacy, resulting from the method of grouping." He concluded that Secrist's results "prove nothing more than that the ratios in question have a tendency to wander about."[7]

It is remarkable, especially considering how old and well known regression effects are, how often these effects are mistaken for something substantive. Although Secrist himself was a professor of statistics, Willford I. King, who in 1934 wrote a glowing review of Secrist's book, was president of the American Statistical Association![8] This error was repeated in a book by W. F. Sharpe, a Nobel laureate in economics, who ascribed the same regression effect Secrist described to economic forces. His explanation of the convergence, between 1966 and 1980, of the most profitable and least profitable companies was that "ultimately economic forces will force the convergence of profitability and growth rates of different firms."[9] The explanation is statistical, not economic. Apparently, this led Milton Friedman, yet another Nobel laureate in economics, to try to set his colleagues straight.[10]

In 1927, Truman Kelley described a specific instance of a regression formula of great importance in many fields, although it was proposed for use in educational testing.[11] It shows how you can estimate an examinee's true score from his/her observed score on a test. *True score* is the psychometric term for the average of the person's observed scores if they took essentially identical tests over and over again forever.* Kelley's equation relates

* *Essentially identical tests* is shorthand for what psychometricians call "parallel forms" of the test. This means tests that are constructed of different questions but span the same areas of knowledge, are equally difficult, and are equally well put together. In fact, one part of the formal definition is the notion that, if two tests were truly parallel, a potential examinee would be completely indifferent as to which form was actually administered.

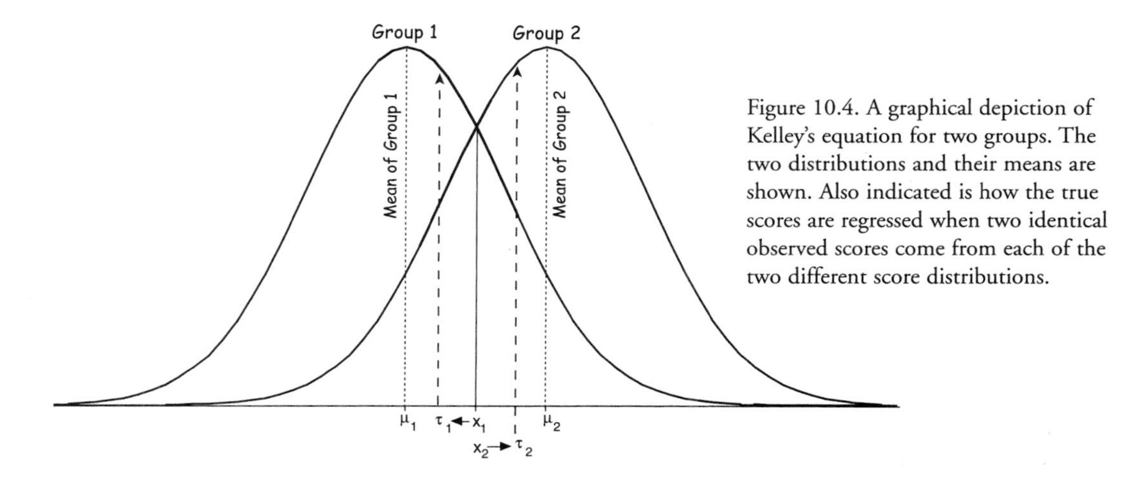

Figure 10.4. A graphical depiction of Kelley's equation for two groups. The two distributions and their means are shown. Also indicated is how the true scores are regressed when two identical observed scores come from each of the two different score distributions.

the estimated true score ($\hat{\tau}$) to the observed score (x). It tells us that the best estimate is obtained by regressing the observed score in the direction of the mean score (μ) of the group that the examinee came from. The amount of the regression is determined by the reliability (ρ) of the test. Kelley's equation is

$$\hat{\tau} = \rho x + (1 - \rho)\,\mu \qquad (1)$$

Note how Kelley's equation works. If a test is completely unreliable ($\rho = 0$), as would be the case if each examinee's score were just a random number, the observed score would not count at all and the estimated true score would be merely the group mean. If the test scores were perfectly reliable ($\rho = 1$), there would be no regression effect at all, and the true score would be the same as the observed score. The reliability of virtually all tests lies between these two extremes and so the estimated true score will be somewhere between the observed score and the mean.

A diagram aids intuition about how Kelley's equation works when there are multiple groups. Shown in figure 10.4 are the distributions of scores for two groups of individuals, here called Group 1 (lower-scoring group) and Group 2 (higher-scoring group). If we observed a score x, midway between the means of the two groups, the best estimate of the true score of the individual who generated that score depends on which group that person belongs to. If that person came from Group 1, we should regress the score downward; if from Group 2, we should regress it upward.

The regression effect occurs because we know that there is some error in the score. The average error is defined to be zero, and so some errors will be positive and some negative. Thus, if someone from a low-scoring group has a high score, we can believe that to some extent that person is the recipient of some positive error, which is not likely to reappear upon retesting, and so we regress that score downward. Similarly, if someone from a high-scoring group has an unusually low score, we regress that score upward.

So far this is merely an equation. What is the paradox? Webster defines a paradox as a statement that is opposed to common sense and yet is true. So long as Kelley's equation deals solely with abstract groups numbered

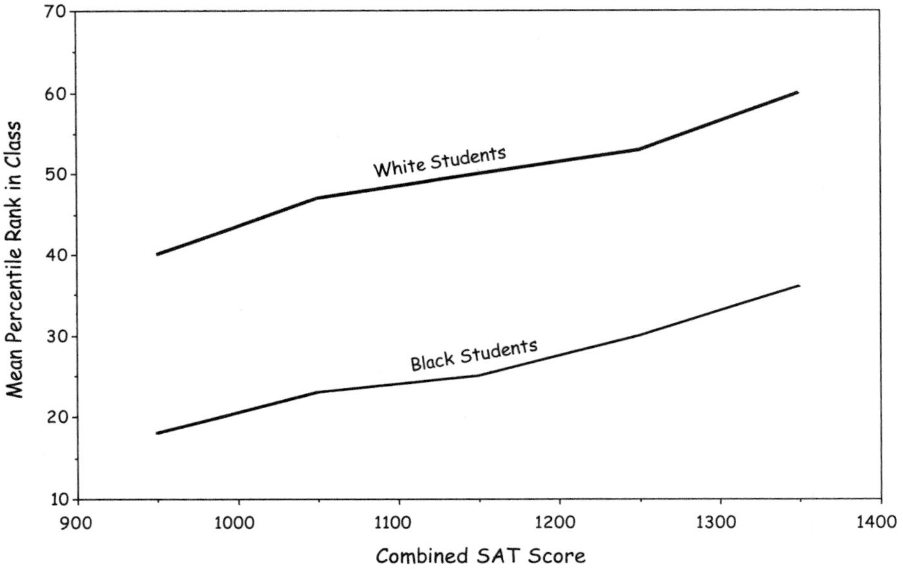

Figure 10.5. An accurate revision of figure 3.10 from Bowen and Bok (1998) showing that at all levels of SAT score, black students' performance in college courses is much lower than white students with matched SAT scores. This bears out the prediction made by Kelley's equation.

1 and 2, no paradox emerges. But suppose we call Group 1 the "Low SES Group" and Group 2 the "High SES Group."* Now when we see someone from Group 1 with a high score, despite their coming from an environment of intellectual and material deprivation, we suspect that they must be very talented indeed and their true ability ought to be considered somewhat higher. Similarly, seeing someone who comes from a more privileged background but scores low leads us to suspect a lack of talent and hence to believe that this person ought to be rated lower still.

Do people truly make this sort of mistake? In the August 31, 1999, issue of the *Wall Street Journal*, an article described a research project, done under the auspices of the Educational Testing Service, called "Strivers." The goal of "Strivers" was to aid colleges in identifying applicants (usually minority applicants) who have a better chance of succeeding in college than their test scores and high school grades might otherwise suggest. The basic idea was to predict a student's SAT score from a set of background variables (e.g., ethnicity, SES, mother's education) and characterize some of those students who do much better than their predicted value as "Strivers." These students might then become special targets for college admissions officers. In the newspaper interview, the project's director, Anthony Carnevale, said, "When you look at a Striver who gets a score of 1000, you're looking at someone who really performs at 1200." Harvard emeritus professor Nathan Glazer, in an article on Strivers in the September 27, 1999, *New Republic*, indicated that he shares this point of view when he wrote (p. 28), "It stands to reason that a student from a materially and educationally impoverished environment who does fairly well on the SAT and better than other students who come from a similar environment is probably stronger than the unadjusted score indicates."

Unfortunately for the goals of affirmative

* SES is a commonly used sociologist acronym meaning "socioeconomic status"; low SES is generally interpreted as pertaining to poor people who are members of minorities.

action, neither Mr. Carnevale nor Mr. Glazer is correct. When you look at a Striver who gets a score of 1,000, you are probably looking at someone who really performs at 950. And, alas, a Striver is probably *weaker* than the unadjusted score indicates.

A Somewhat Technical Obiter Dictum

This result is distressing for those who believe that affirmative action is justified because of their mistaken belief that standard admission test scores underpredict the subsequent performance of low-SES students. Exactly the opposite is true, for ample evidence has shown that students who are admitted with lower than usual credentials do worse than expected, on average. For example, in their exhaustive study of the value of affirmative action, William Bowen and Derek Bok produce a figure (figure 10.5) that shows that black students' rank in college class is approximately 25 percentile points lower than white students with the same SAT scores. In this metric, the effect is essentially constant over all SAT scores, even the highest.

In a more thorough analysis, Ramist, Lewis, and McCamley-Jenkins (1994) used data from more than forty-six thousand students, gathered at thirty-eight colleges and universities, to build a prediction model of college performance based on precollegiate information. One analysis (among many) predicted first-year college grade point average from SAT score, from high school grade point average (HS-GPA), and from the two combined. They then recorded the extent to which each ethnic group was over- or underpredicted by the model. An extract of their results is shown in table 10.5. The metric reported is grade points, so a one-point difference corresponds to one grade level (a B to a C, for example). The entries in this table indicate the extent to which students' first-year grade point averages were under- or overpredicted by the model based on the variables indicated. Thus the entry "0.02" for Asian American students for a prediction based on just their high school grades means that Asian Americans actually did very slightly (0.02 of a grade level) better than their high school grades predicted. And the "–0.35" for black students means that they did about a third of a point worse than their grades predicted. Note that while SAT scores also overpredict minority students' college performance, the extent of the error they make is somewhat smaller.

Table 10.5
Prediction Model of Collegiate Performance from Precollegiate Information

	Ethnic group			
Predictor	Asian American	White	Hispanic	Black
HS-GPA	0.02	0.03	-0.24	-0.35
SAT	0.08	0.01	-0.13	-0.23
HS-GPA and SAT	0.04	0.01	-0.13	-0.16
Sample Sizes	3,848	36,743	1,599	2,475

Source: Data from Ramist, Lewis, and McCamley-Jenkins (1994).

The results are very clear: the common regression model overpredicts the performance of minority populations. This result matches the regression phenomenon I described as Kelley's paradox. And it matches it in amount as well as direction. These results are not the outcome of some recent social change, but have been observed for decades.[12]

The conclusions I drew from what is predicted by Kelley's equations are borne out repeatedly by the evidence.

11 Order in the Court

Because we are almost never interested in seeing Alabama first, it is astonishing how often data displays use alphabetical order as the organizing principle of choice. The only reason I can think of is that organizing by this trivial aspect of the data is easy and obvious. Yet the benefits derived from ordering a display alphabetically are usually so modest and the gains from reordering so large that one would expect wiser alternatives to prevail. But typically they do not.*

Consider the function shown in figure 11.1 that relates various foods to their cost. The order of the foods along the horizontal axis is alphabetical. Yet the same data, displayed for a French audience (figure 11.2) is quite different. Do we want to use any sort of organizing principle that changes the viewer's perceptions so radically, when the change in the data is superficial? For more mathematically inclined readers, this is an example of a function that is not invariant with respect to translation.

Usually the effectiveness of a good display increases with the complexity of the data. When there are only a few points, almost anything will do; even a pie chart with only three

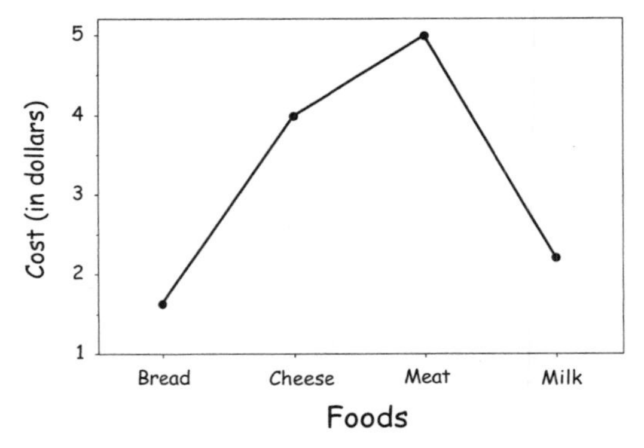

Figure 11.1. A simple line plot showing the price of four foods. The foods are ordered alphabetically.

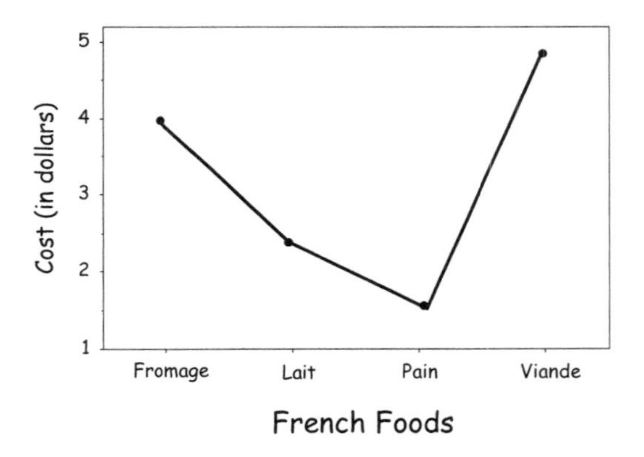

Figure 11.2. The same four foods at the same prices, still ordered alphabetically; but because their names are now in French, the order is different and hence the picture looks different.

I am grateful to Karen McQuillen for tracking down the arms importation data and to Jeff Smith for telling me about Jacob Bronowski's insight.

* A sensible rule for deciding on the way to order a display is "order by those aspects of the data in the table that are of interest." We are only rarely interested in "Alabama first!" But for one instance where alphabetical order is helpful, see figure 13.6.

or four categories is usually comprehensible. Indeed, this is one explanation for the empirical finding of so many dreadful displays of data—such displays may have been minimally satisfactory for small data sets, and so the same format was used again for a larger data set of the same type. Thus my goal of changing practice would be helped if I could find a small example that yields insights into the underlying structure when well displayed, but those insights are effectively hidden with a less carefully chosen format. It would also help if the example were interesting.

My prayers were answered by a plot in the "News of the Week in Review" section of the *New York Times* (July 2, 2000, p. 5), which contained information about how the justices of the U.S. Supreme Court voted on six important cases. I will not reproduce that plot here but instead provide my version, figure 11.3, which captures the key elements of the original.

The justices (the columns) and the topics of the cases are ordered alphabetically. The principal feature that we can make out is that Sandra Day O'Connor always voted in the majority. Other than that, nothing jumps out. Reordering both rows and columns in an

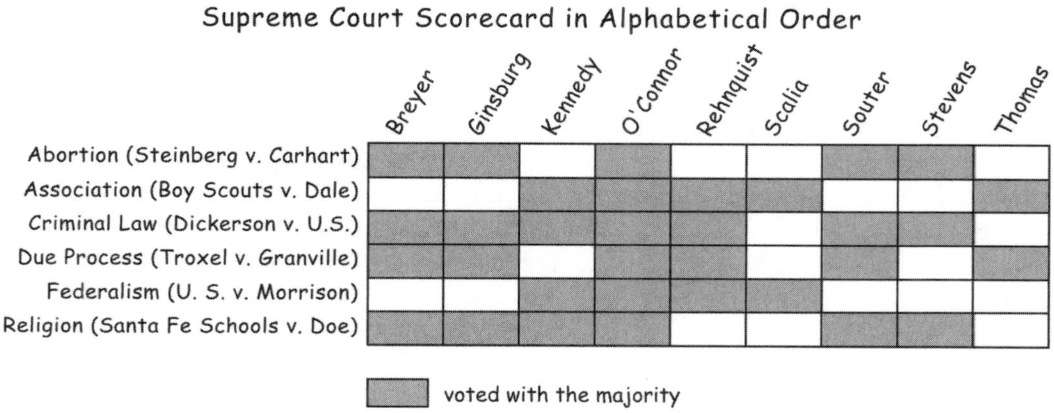

Figure 11.3. A graphical depiction of the voting of the U.S. Supreme Court on six important cases (from *New York Times*, "News of the Week in Review," July 2, 2000, p. 5), in which both the cases and the justices are ordered alphabetically.

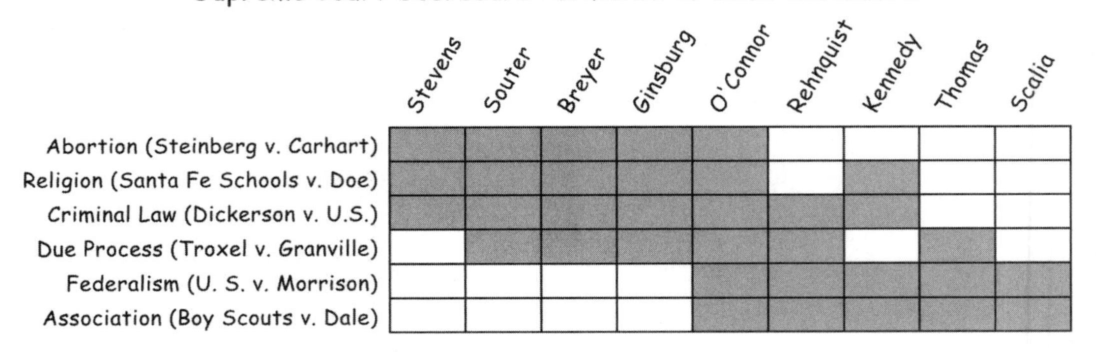

Figure 11.4. A reordering of both cases and justices that makes clear the ideological structure of the court, at least as portrayed in these cases.

obvious, but data-related, manner tells a wonderful story (see figure 11.4). We can easily identify the liberal bloc of the court (Souter, Breyer, and Ginsburg), who always voted together and are the orthogonal complement of Justice Scalia, the right-most justice. O'Connor's central position is now represented both metaphorically and geometrically. And Justices Thomas and Stevens don't seem to share too many opinions.

What is remarkable in this example is not that a good display can expose structure that a poor one obscures, but rather how much a difference can be made with as simple a data set as this and how little effort it took to turn a bad display into a good one. Their format is identical; only their organization—the ordering of the rows and columns—differs. Why does ordering make a difference? What is it communicating? A revealing order of the data is one that reflects an underlying dimension of the data that is reflected in the data values. The ordering makes that latent dimension explicit. Although nothing is explicitly stated about the political orientation of the Supreme Court justices in the data shown in figure 11.4, the order that is manifested by their behavior is informative. It also tells us that if these cases are representative of the mass of cases heard by the court, Justice O'Connor is the one to watch if you are interested in predicting the outcome.

If ordering can make such a difference with a modest-sized data set, when there are only fifty-four binary data points, should we expect a real bonanza of insight from reconfiguring a larger data set? It depends. It depends on the character of the structure that lies within the larger data set. Even the finest display cannot extract meaning that is not there in the first place. The special value of good display (in this instance, wise ordering) with a larger data set resides in the difficulty of extracting meaning and structure from a large data set without such cleverness. One could argue that an astute observer would uncover the character of the Supreme Court even from the ordering in figure 11.3; but as displays increase in complexity, the likelihood of such a happy result diminishes (probably as the square of the dataset size).

Let us consider another, somewhat larger example. Table 11.1 shows the world's ten biggest suppliers of arms and the fifty biggest recipients of arms for the years 1975–1979. The entries in the table are the dollar value of

Table 11.1
Value of Arms Transfers, Cumulative 1975–1979, by Major Supplier and Recipient Country (in millions of current dollars)

	Suppliers											
Recipients	Total	Soviet Union	United States	France	United Kingdom	Germany	Czecho-slovakia	Italy	Poland	China	Canada	Others
Afghanistan	**465**	450	–	–	–	–	10	–	–	–	–	5
Algeria	**1,940**	1,500	–	10	–	350	–	10	–	–	–	70
Angola	**845**	500	–	5	10	10	10	–	20	–	–	290
Argentina	**970**	–	90	270	60	110	–	80	–	–	–	360
Australia	**690**	–	420	–	130	130	–	–	–	–	–	10
Belgium	**615**	–	270	40	120	140	–	–	–	–	5	40
Brazil	**740**	–	160	50	400	20	–	80	–	–	–	30
Bulgaria	**1,230**	1,200	–	–	–	–	20	–	10	–	–	–
Canada	**1,035**	–	825	–	10	190	–	–	–	–	–	10
China	**630**	210	–	50	350	–	–	–	–	–	10	10
Cuba	**875**	875	–	–	–	–	–	–	–	–	–	–
Czechoslovakia	**1,250**	1,200	–	–	–	–	–	–	–	–	–	50
Ecuador	**575**	–	40	280	70	110	–	5	–	–	10	60
Egypt	**1,500**	250	250	490	110	180	20	60	–	60	–	80
Ethiopia	**1,830**	1,500	90	10	–	5	30	20	10	5	–	160
Germany, East	**2,005**	1,700	–	–	–	–	230	–	70	–	–	5
Germany, West	**2,230**	–	1,600	140	10	–	–	10	–	–	10	460
Greece	**1,990**	–	1,200	390	20	230	–	60	–	–	10	80
Hungary	**1,035**	975	–	–	–	–	50	–	10	–	–	–
India	**2,150**	1,800	40	40	100	10	50	20	40	–	–	50
Iran	**8,770**	650	6,600	200	310	430	–	340	–	–	–	240
Iraq	**6,780**	4,900	–	410	20	160	80	70	30	10	–	1,100
Israel	**4,300**	–	4,200	10	60	–	–	30	–	–	–	–
Italy	**620**	–	550	–	–	10	–	–	–	–	–	60
Japan	**745**	–	725	–	20	–	–	–	–	–	–	–
Jordan	**590**	–	500	–	20	5	–	–	–	–	5	60
Korea, North	**580**	280	–	–	–	10	–	–	10	170	–	110
Korea, South	**1,925**	–	1,700	10	5	80	–	50	–	–	40	40
Kuwait	**790**	50	350	150	210	20	–	–	–	–	–	10
Libya	**6,910**	5,000	–	310	10	160	270	450	250	–	–	460
Morocco	**1,415**	20	310	725	5	50	5	50	–	–	–	250
Netherlands	**690**	–	370	–	120	80	–	10	–	–	10	100
Pakistan	**895**	20	180	320	20	–	5	–	–	240	–	110
Peru	**1,090**	650	100	110	10	40	–	80	–	–	–	100
Poland	**1,420**	1,200	–	–	–	–	210	–	–	–	–	10
Romania	**875**	675	–	70	–	–	30	–	70	10	–	20
Saudi Arabia	**3,590**	–	1,800	290	900	20	–	130	–	–	–	450
South Africa	**535**	–	20	310	–	–	–	50	–	–	5	150
Soviet Union	**2,850**	–	–	–	–	–	1,900	–	800	–	–	150
Spain	**1,020**	–	550	170	–	90	–	20	–	–	10	180
Switzerland	**580**	–	460	–	10	–	–	–	–	–	10	100
Syria	**4,500**	3,600	–	190	30	100	310	–	10	–	–	260
Turkey	**1,110**	–	550	–	–	290	–	240	–	–	–	30
United Kingdom	**915**	–	825	30	–	–	–	–	–	–	–	60
United States	**790**	–	–	5	230	70	–	5	–	–	250	230
Viet Nam, North	**1,320**	1,300	–	–	–	–	–	–	–	10	–	10
Viet Nam, South	**850**	–	850	–	–	–	–	–	–	–	–	–
Yemen (Aden)	**580**	575	–	–	–	–	–	–	–	–	–	5
Yemen (Sanaa)	**620**	210	110	80	–	5	–	5	100	–	–	110
Yugoslavia	**630**	525	10	20	10	–	–	10	5	–	20	30
Total	**82,885**	**31,815**	**25,745**	**5,185**	**3,380**	**3,105**	**3,230**	**1,885**	**1,435**	**505**	**395**	**6,205**

Table 11.2
Value of Arms Transfers, Cumulative 1975–1979, by Major Supplier and Recipient Country, Sorted to Show Structure, Some Middle Eastern Recipients Shaded for Emphasis (in millions of current dollars)

	Suppliers								
Recipients	**Soviet Union**	Czechoslovakia	Poland	**United States**	United Kingdom	France	West Germany	Italy	Others
Libya	**5,000**	270	250			310	160	450	460
Iraq	**4,900**					410	160		1,100
Syria	**3,600**	310				190	100		260
India	**1,800**				100				
Germany, East	**1,700**	230							
Ethiopia	**1,500**								165
Algeria	**1,500**						350		
Viet Nam, North	**1,300**								
Bulgaria	**1,200**								
Czechoslovakia	**1,200**								
Poland	**1,200**	210							
Hungary	**975**								
Cuba	**875**								
Romania	**675**								
Peru	**650**			**100**		110			100
Yemen (Aden)	**575**								
Yugoslavia	**525**								
Angola	**500**								290
Afghanistan	**450**								
Korea, North	**280**								280
South Africa						310			155
Ecuador						280	110		
Argentina						270	110		360
China	**210**				350				
Yemen (Sanaa)	**210**		100	**110**					110
Brazil				**160**	400				
Pakistan				**180**		320			350
Egypt	**250**			**250**	110	490	180		140
Belgium				**270**	120		140		
Morocco				**310**		725			250
Kuwait				**350**	210	150			
Netherlands				**370**	120				110
Australia				**420**	130		130		
Switzerland				**460**					110
Jordan				**500**					
Turkey				**550**			290	240	
Spain				**550**		170			190
Italy				**550**					
Japan				**725**					
Canada				**825**			190		
United Kingdom				**825**					
Viet Nam, South				**850**					
Greece				**1,200**		390	230		
Germany, West				**1,600**		140			470
Korea, South				**1,700**					
Saudi Arabia				**1,800**	900	290		130	450
Israel				**4,200**					
Iran	**650**			**6,600**	310	200	430	340	240
Total	**31,725**	**1,020**	**350**	**25,455**	**2,750**	**4,755**	**2,580**	**1,160**	**5,600**

arms that each recipient received from each supplier (in millions). The table is abstracted from a much larger version in the *World Military Expenditures and Arms Transfers, 1970–1979*, which ordered the table alphabetically within continent. I have omitted about 70 percent of the recipient countries (to conserve space), keeping only the fifty recipients that received the largest amounts. The original table had fifty African countries, ordered alphabetically, so my version here, which orders these fifty alphabetically, is not doing the designers of the original table a serious injustice.

Once again, I reordered both rows and columns; the rows by the amounts received from either the Soviet Union or the United States, and the columns by similarity to either the Soviet Union or the United States. Next, I omitted any amounts less than $100 million (to get rid of visual clutter), and an interesting pattern emerged (see table 11.2). We immediately see the division of the world into two parts; one part armed by the United States and its allies, and the other by the Soviet Union and its allies. But the data entries that violate this rule are especially interesting. Clearly, the money available from those Middle Eastern countries that fell within the Soviet sphere was too much to resist for France, Germany, and Italy. This result suggests where France's current pro-Iraq policies may have their origin

The format of table 11.1 surely is a better storage medium, allowing us to find details about specific countries easily, but it provides no glimpse of the deep structure that is so readily visible in the reformatted version (table 11.2).

This gets me back to the original question. Why do those data display preparers in the mass media, whose primary goal is surely communication of quantitative phenomena and not efficient data storage, so often retreat to "Alabama first"? I do not doubt that they wish to provide us with the truth, but they certainly have a peculiar way of conveying it. It is reminiscent of Jacob Bronowski's wonderful observation on the Spanish Inquisition: "They cared passionately for the truth, but their sense of evidence was different from ours."[1]

12 No Order in the Court

In chapter 11 we illustrated why we are not usually interested in Alabama first. Alphabetical order may be important for some limited set of purposes, but much more often another ordering is likely to be far more useful. The examples in chapter 11 demonstrated how reordering helps us to understand the deep structure that may underlie the data in the display. Sometimes it also helps us to see exactly what is in the display.

The benefits associated with the proper ordering arose dramatically on March 17, 1998, in a murder trial, when *State v. Gibbs* began in Connecticut Superior Court. The trial was held in the Hartford courthouse of the Hartford—New Britain judicial district. The public defender's office used the case to challenge Connecticut jury arrays, alleging that they do not include a sufficient number of Hispanics. One aspect of the case was the assertion that court summonses are not delivered as frequently to Hispanics as to others.

To support this allegation, the public defender introduced a graph (Defendant's Exhibit X-15, shown here as figure 12.1). The horizontal axis lists census tract numbers. The vertical axis is percentage. The dashed line connects points that represent the percentage of Hispanics in each tract. The solid lines connect points that represent the percentage of undeliverable (by mail) summonses in those tracts for each of five court years.

The goal of this display was to reveal the extent to which there is a positive relationship between undelivered court summonses for jury duty and the size of the Hispanic population. Other than the impression of chaos, all we seem to see is that there is some consistency, across time, of the likelihood of a court summons being delivered in any particular census tract. If there is any relation to the size of the Hispanic population, I cannot see it.

The reason why this graph fails is clear: the horizontal axis is being wasted carrying the irrelevant information of census tract number. Had that axis been used for one of the variables of interest, the result would have supported the defense's position. Unlikely as it might seem, such a plot was prepared by Stephen Michelson, a statistical analysis consultant and president of Longbranch Research Associates, who was engaged by the state's attorney's office, the prosecutors in the case. Mr. Michelson prepared separate plots for each year. His figure for 1993—1994 is reproduced here as figure 12.2. The horizontal axis

I am grateful to Stephen Michelson for sharing this interesting example with me. It suggests that the task of graphical education is not over. Paraphrasing Einstein's famous remark, "old graphical methods don't die, just the people who practice them."

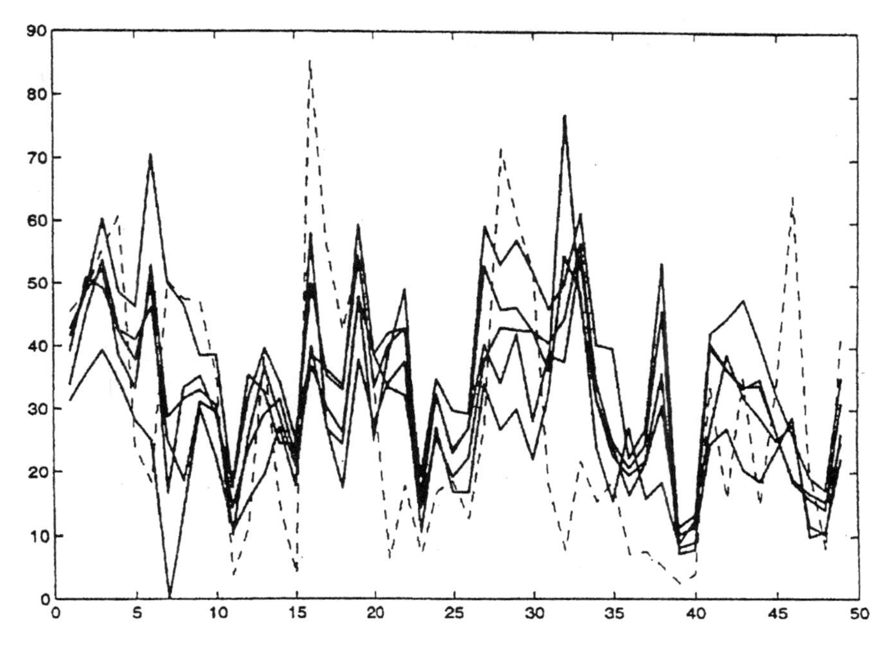

Figure 12.1. Percent undeliverable summonses in Hartford census tracts. Tract numbers run from 5,001 through 5,049 on the horizontal axis (leading digits omitted for clarity). The dotted line shows percentage Hispanic in the over-nineteen population of each tract. The five solid lines connect the percentages of undelivered (code 13) summonses for each of the court years.

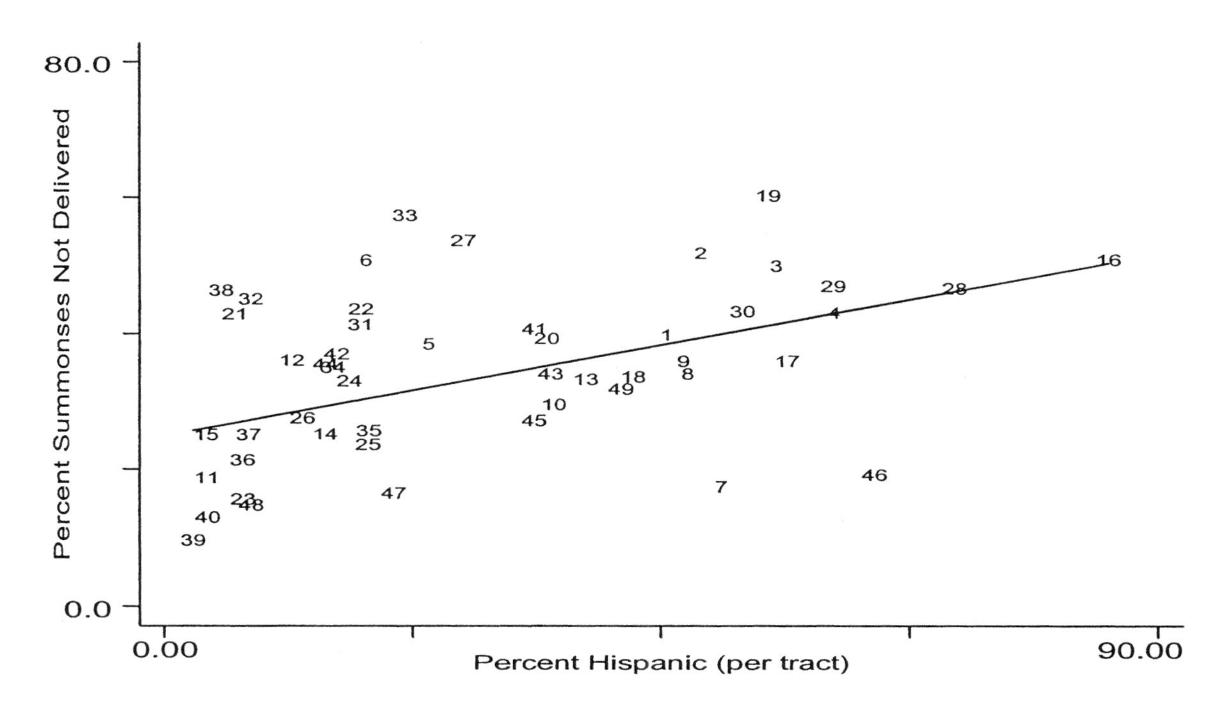

Figure 12.2. The rate of undelivered summonses in forty-nine census tracts shown as a function of percent Hispanic in those census tracts for 1993–1994. (Plot prepared by Stephen Michelson.)

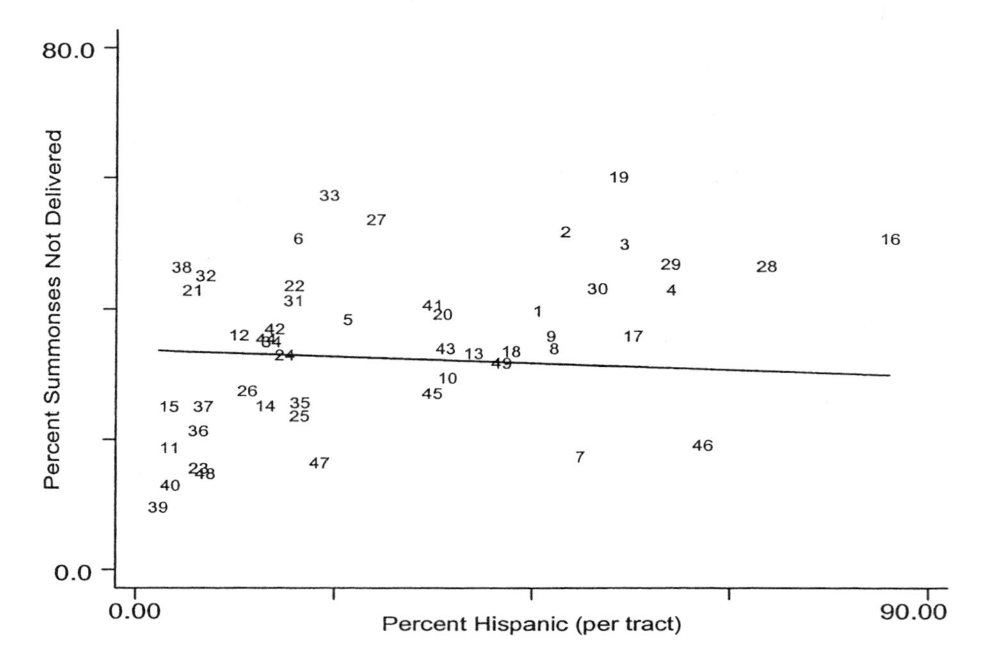

Figure 12.3. The adjusted rate of undelivered summonses in forty-nine census tracts shown as a function of percent Hispanic in those census tracts for 1993–1994. (Plot prepared by Stephen Michelson.)

is the percent Hispanic in each district, and the vertical axis the percentage of summonses not delivered. The census tract number is used as the plotting symbol. The positive relationship, albeit weak, is obvious and indicated by the drawn-in regression line.

The key to the success of Michelson's plot is that the data are ordered by an aspect that matters rather than something arbitrary.

Although not germane to the principal point of this chapter, it is natural to be curious about why the prosecution would re-present the defense's exhibit in a way that makes the defense's point more explicitly. A good question. I suspect that the prosecution needed a good display of the data to be able to make their point, which was that after adjustment for other economic variables, the apparent underrepresentation of Hispanics on Connecticut's juries disappears. In fact, after such an adjustment, one discovers a small but statistically significant effect of more summonses delivered as the proportion of Hispanics in a census tract increases (see figure 12.3). This yields an overrepresentation of Hispanics on Connecticut's juries, exactly the opposite of what the defense alleged.

Although many technical issues surrounded the use of covariates and the interpretation of proxy variables, this argument was apparently accepted, for the court found in favor of the prosecution.

13 Like a Trout in the Milk

During a dairymen's strike in New England, some suspected that the limited supplies of milk were being watered down for wider distribution. Commenting on the worth of the evidence being cited, Henry David Thoreau wrote in his journal (November 11, 1850): "Sometimes circumstantial evidence can be quite convincing; like when you find a trout in the milk."

Oftentimes a statistical graphic provides the evidence for a plausible story, and the evidence, though perhaps only circumstantial, can be quite convincing. For example, in figure 13.1 we cannot help but note the obvious causal implications in the simple swooping line that bemoans the fate of the bald eagle prior to the banning of DDT in 1973 and the miraculous recovery afterward. As with Thoreau's "trout in the milk," the evidence may be circumstantial but is convincing.

But such graphical arguments are not always valid. Knowledge of the underlying phenomena and additional facts may be required. In figure 13.2 we see that men have been making slow but steady progress, over

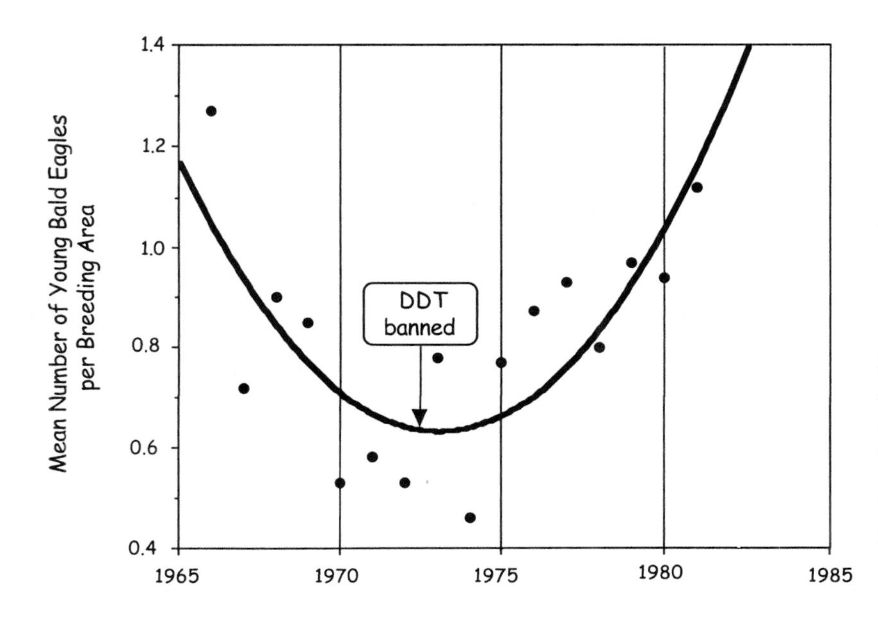

Figure 13.1. A plot showing the average number of bald eagle hatchlings per breeding area over the sixteen-year period 1966 through 1981. The banning of DDT in 1973 led to the eagles' resurgence. Data from Cleveland (1994, p. 31).

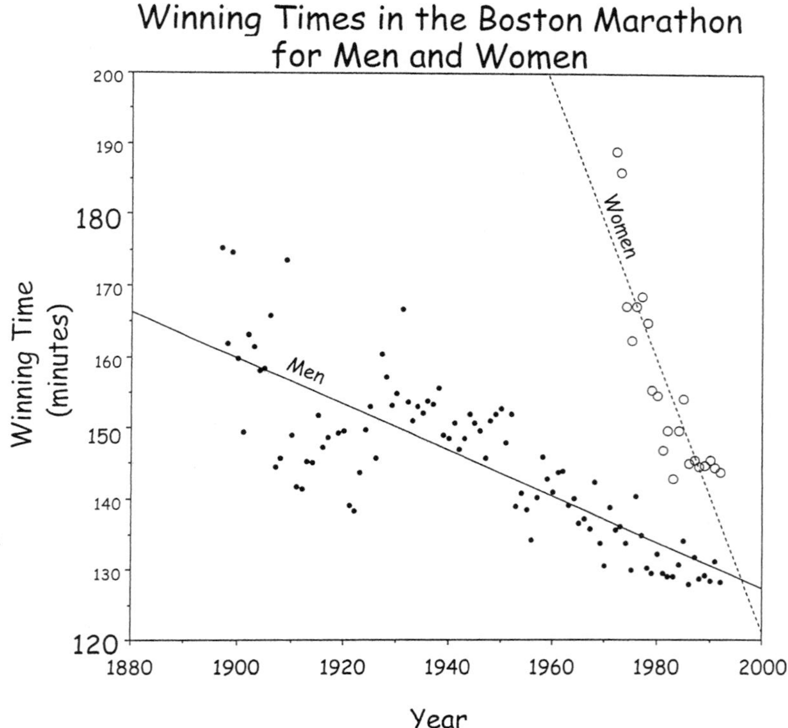

Figure 13.2. Winning times for the Boston Marathon for men (1897 through 1992) and women (1972 through 1992), augmented with the best linear fit. Data from Meserole (1992, pp. 619–620).

the last century or so, toward an eventual two-hour time for the Boston Marathon. Women, on the other hand, have been making very rapid progress since they were first allowed to run the Boston Marathon in 1972. From the graph, any fool would have predicted that women would be beating men and be very close to the two-hour marathon by the year 2000. Such is the power of this graphical display that some misguided sports reporters have actually made this inference.*

The preceding two examples show situations in which a simple causal inference can be drawn from the graphic. In both of them, the

* This example was first suggested to me by Andrew Gelman (Gelman and Nolan 2002, page 20).

Such linear extrapolation is not always as dopey an idea as it might first appear; after all, all functions are pretty linear over a short enough segment. Consider the world record times for the mile over the first half of the twentieth century shown in figure 13.3. We see consistent linear improvements of a bit over one-third of a second per year that were contributed to by legendary milers such as the American Glenn Cunningham, the "Flying Finn" Paavo Nurmi, and the other fine Scandinavian runners.

If we use this fitted straight line to predict the world record in the year 2000 we would find (figure 13.4), remarkably, that the world record for the mile has continued to improve at about the same rate. Indeed, the pace of improvement has even increased a bit. The next forty years of this improvement has been the work of the Brits. Roger Bannister was the first sub-four-minute miler, but he was followed quickly by the Australian John Landy, who in turn was joined by his countryman Herb Elliott and then the great New Zealand miler Peter Snell. There was a ten-year interlude when the American Jim Ryun, while barely out of high school, lowered the mark still further, before losing the record to Filbert Bayi, the first African to hold it. Then John Walker, Sebastian Coe, and Steve Ovett reestablished Britain's domination of the event.

The decade of the 1990s marked the entrance of North Africans into the competition as first Noureddine Morceli of Algeria lowered the record and then the Moroccan Hicham El Guerrouj brought it to its current point of 3:43.13.

Obviously this linear trend cannot continue forever, but what about for another fifty years? (See figure 13.5.) It seems unlikely that in 2050 the record for the mile would be as fast as 3:21, but what would we have said in 1950 about a 3:43 mile? Thus the question is not about the linearity of improvement over a short period of time, but rather about how long "short" is.

evidence for the inference is circumstantial, but (for some audiences at least) quite convincing. In neither case is there a long narrative associated with the display, although a deeper look might provide one (e.g., what is behind the dip in men's times between 1900 and 1920?). The graphic as a narrative form has a rich heritage, and the best example is surely Minard's plot of Napoleon's Russian Campaign.* My next example, although in content far less heart-rending than Minard's visual memorial of four hundred thousand French soldiers, shows how a simple graphic can provide convincing, albeit circumstantial, evidence to support a rich narrative.

A Story of Faxes about Taxes

Every year, Canada's federal government releases a budget. The budget's contents are a secret until the minister of finance rises in the House of Commons, usually at 4:30 p.m., and releases it.

Among those who pay close attention to its contents are tax professionals in law and accounting firms. The major accounting firms (including Price Waterhouse) have a tradition

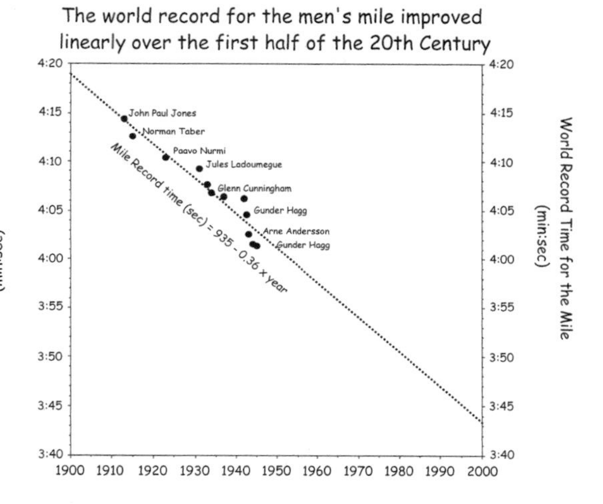

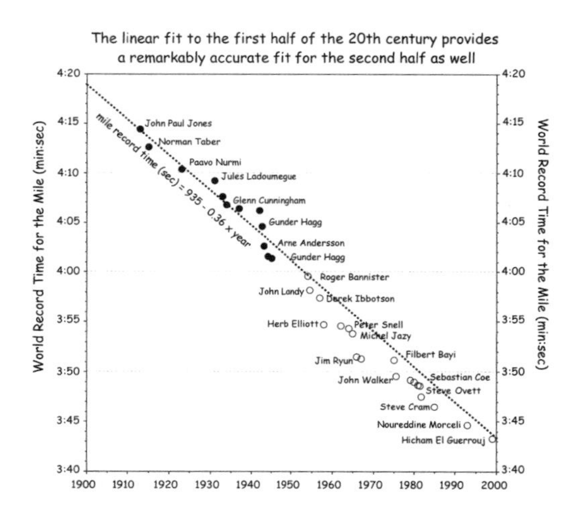

Figure 13.3 (left). The world records for the mile from 1900 until 1950 shown with the best-fitting linear function.

Figure 13.4 (bottom left). The world records for the mile from 1900 until 2000 shown with the best-fitting linear function from the first fifty years.

Figure 13.5 (below). The world records for the mile from 1900 until 2000 shown with the best-fitting linear function from the entire century, indicating extrapolated values for the records over the next fifty years.

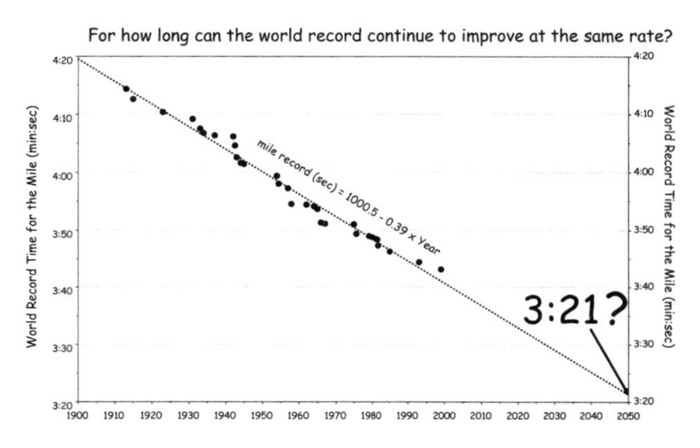

* See Tufte 1983 or Wainer 1984, 1994, or 1997b for a reproduction of this wonderful data display and an associated description.

of putting out bulletins for clients and contacts, concentrating mainly on tax changes. These are distributed by a variety of means, including fax. Hectic activity extends into the early morning hours.

In 2002, Price Waterhouse engaged a new fax service provider, which was given an electronic list of more than 2,300 fax recipients, in an order that reflected their time zones. Generally, those farther east were at the top of the list. Faxing began the following morning.

A few days later, the fax company provided a report. For each intended recipient, it gave the time at which the final attempt to fax was made (whether successful or not), a code indicating the number of attempts that had been made (one, two, three, or four) and either the word *completed* or a short description of the reason for failure.

The report was not in any immediately obvious order.

A variety of graphs were prepared to try to gain some understanding of the process that had taken place. The obvious cumulative plot of completed faxes against time revealed that after the 8:30 start, there was a steady success rate of about seven or eight faxes per minute, for about four hours. The rate then generally dropped gradually, but with a few noticeable bursts at the original rate and a few plateaus at a rate near zero.

Little else was learned until the data were sorted alphabetically and the alphabetical ranking was plotted against the time each fax was received, with a different marker used to indicate whether it got through on the first, second, third, or fourth attempt.

The resulting display is shown as figure 13.6.

From this graph, I inferred the following narrative about the process used:

1. The faxes were sorted alphabetically.
2. They were divided up equally among three transmission systems (A–H, I–N, and O–Z), each of which then went at it at about the same speed, starting at 8:30 a.m. At 10:15, the operator for O–Z went to the bathroom.
3. At noon (or so), all the operators took a break for lunch.
4. A–H was either really lucky and all went through on the first try or had something else to do that afternoon and left. I–N and O–Z were less lucky or more assiduous.
5. After a thirty-minute lunch break, I–N and O–Z shuffled the ones that had not gone through and tried again.
6. Then they picked up the ones that still had not gone through and tried again, and again, and then went home.
7. At 3:00, a second-shift person showed up and found a pile of faxes unsent by A–I (probably they were sent but did not go through and hence were misrecorded) and a few not sent by I–N and O–Z. These were in no particular order, and they were attempted also; after a few tries they were done.

This graph (and much of the description of the situation) was originally prepared by Alan J. Davis, the director of tax communications at Price Waterhouse in Toronto. He concluded (in the understated way that good lawyers often use), "The graph proved useful in discussions with the company about the timing of the faxes." I bet.

Although no one would place Davis's plot of fax deliveries in the same class as Minard's graph of the death of much of a generation of Frenchmen, it is still a remarkable example of the power that well-designed graphical displays can wield in aiding our understanding of large data sets. Of course, the evidence carried in this chart is principally circumstantial, but in this instance at least, we can see the fish splashing.

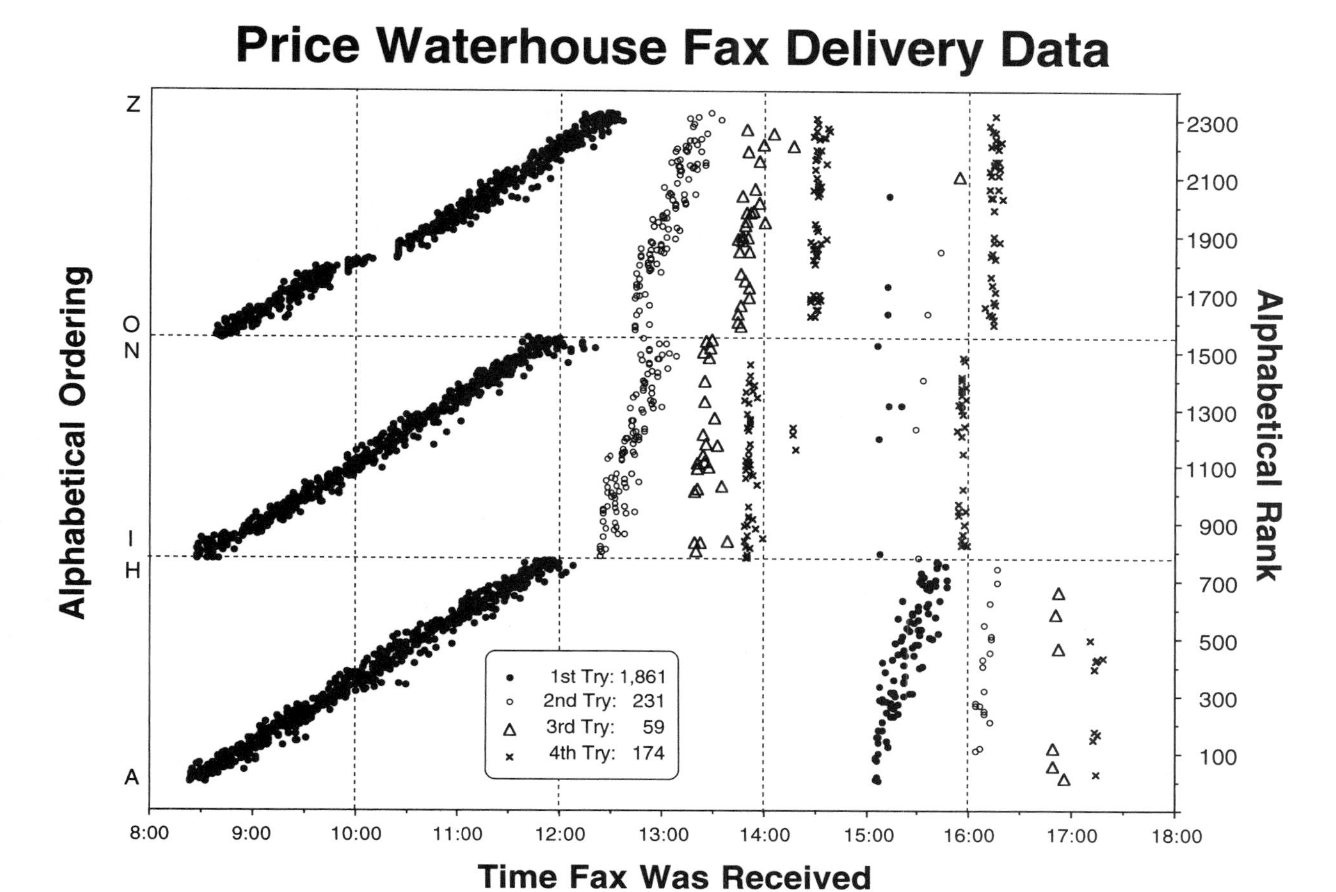

Figure 13.6. A plot of the times that 2,325 faxes were received, shown alphabetically. Data and graphic design from Alan J. Davis (Price Waterhouse–Toronto).

14 Scaling the Market

Placing a fact within a context increases its value greatly. "Our town has two doctors" is a fact, but appending the phrase "but most towns our size have sixteen doctors" tells us so much more. An efficacious way to add context to statistical facts is by embedding them in a graphic. Sometimes the most helpful context is geographical, and shaded maps come to mind as examples. Sometimes the most helpful context is temporal, and time-based line graphs are the obvious choice. But how much time? The ending date (today) is usually clear, but where do you start? The starting point determines the scale. If we start earlier today, the scale is hourly. If we start last year, the scale is monthly, and if we start in 1910, the scale is annual. The starting point and hence the scale are determined by the questions that we expect the graph to answer.

Consider the information about the Dow Jones Industrial Average shown in figure 14.1. It is hard to imagine any useful purpose served by such a plot; the interval being displayed is just too short. What sorts of investment deci-

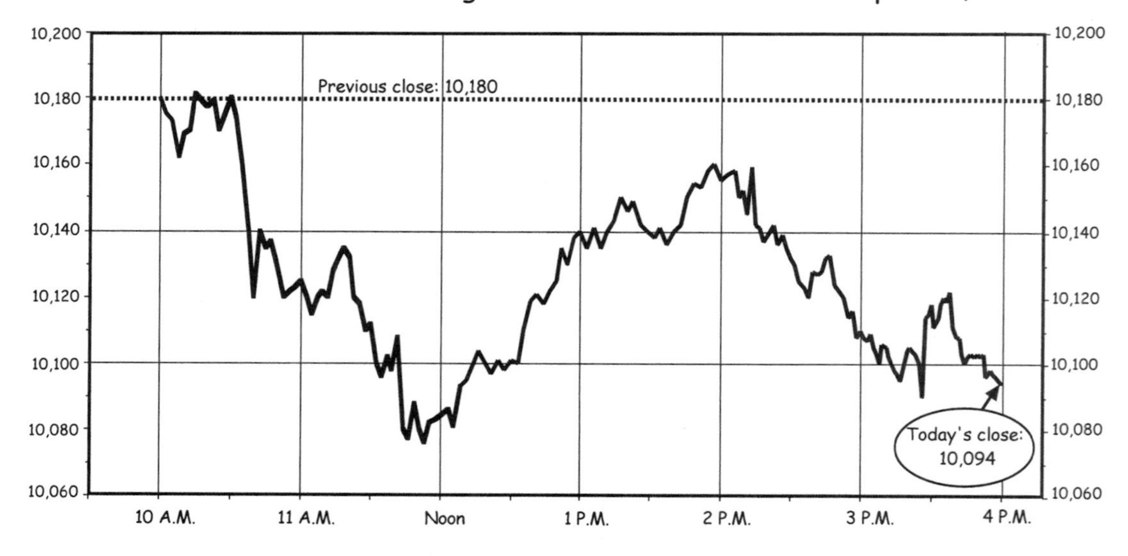

Figure 14.1. The Dow Minute-by-Minute from *New York Times*, March 17, 1998.

sions would be made on the basis of the information in this display? Even for traders who buy and sell the same stock during the course of a single day, what does yesterday's trading pattern tell them about today's? I suspect that the direction of yesterday's movement does not accurately predict today's direction. The only value I can see to such a plot is the prediction of volatility: if yesterday's trading was volatile, then perhaps today's will be, as well.

Of far greater potential use is a plot like that shown in figure 14.2, showing the Dow Jones Average over the course of a year. Investors can tell from such a plot how things have been going and might be able to use this information to make some kinds of intelligent choices in the short term. But for the long-term investor, even this plot is not helpful.

We are often told that buying and holding equities for the long term has been shown historically to be superior to all other investments. A display that shows the exponential growth of three major market indicators over the last eighteen years appears to support this advice (figure 14.3). Although it includes minor perturbations (like the 500-point drop in 1987 and the stock market's reaction to the World Trade Center disaster), it provides strong evidence that you are a fool if you do not put every spare dime into the equity market. Yet this advice is flawed, because even a scale of eighteen years is incomplete and hence insufficient for some kinds of important decisions.

A longer look provides a cautionary note (figure 14.4). If we let the plot range over ninety years, we can see that since 1910, the Dow Jones Average has indeed moved up by two orders of magnitude, a hefty increase by any standard. But in addition we see that this increase has not been uniform. During the eighteen-year period 1964–1982, the market

Figure 14.2. The Dow Jones Industrial Average over the fourteen months from January 1, 2001, until March 1, 2002.

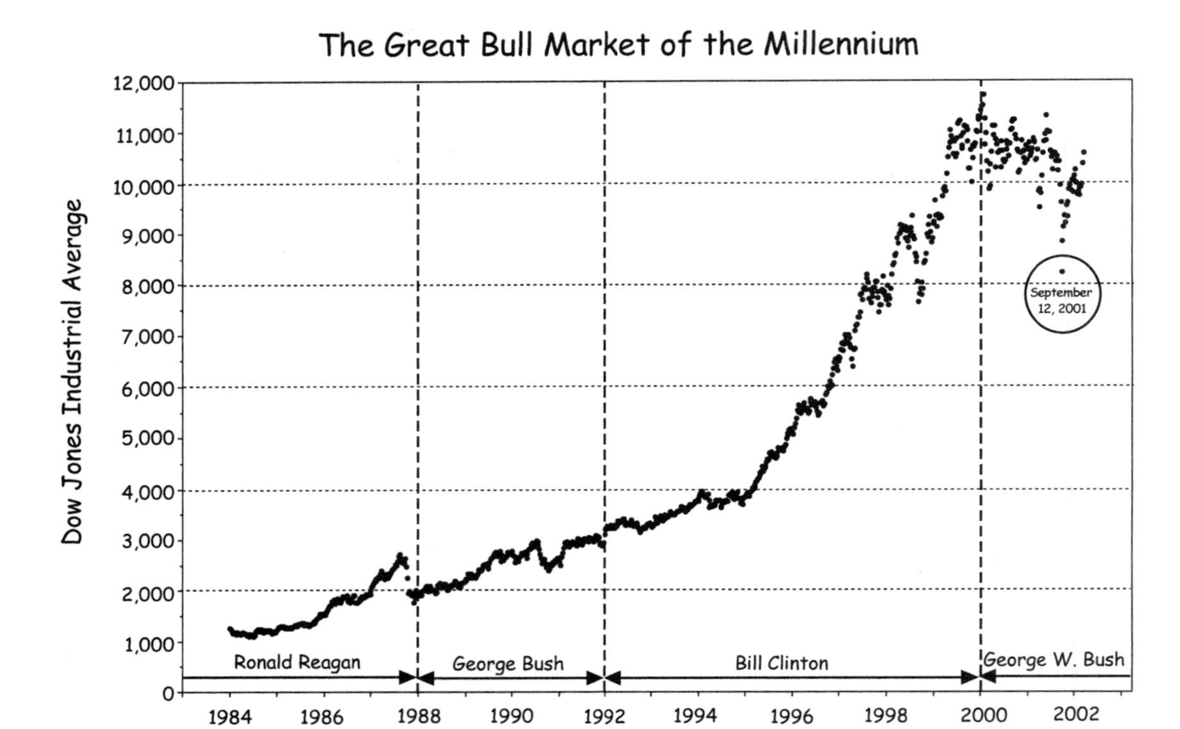

Figure 14.3. The Dow Jones Industrial Average over the past eighteen years, shown weekly.

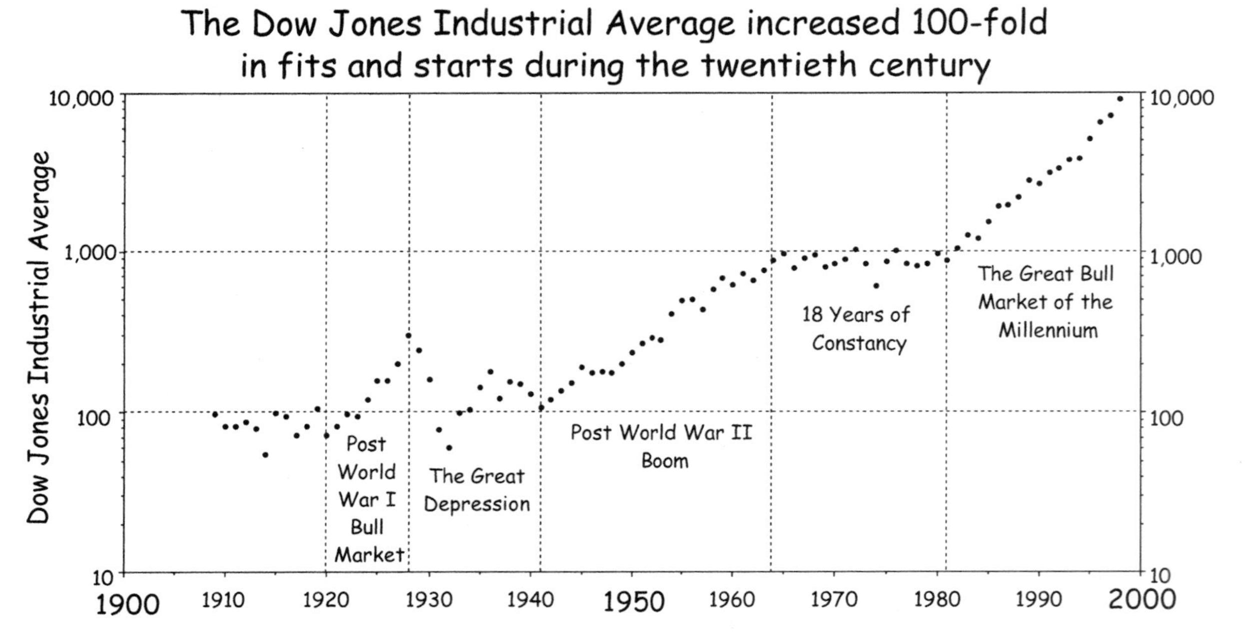

Figure 14.4. The Dow Jones Industrial Average since 1909, shown annually (the value shown for each year is the closing figure on the last day of that year).

was pretty flat, hovering just below 1,000. During this time period, other investments (bonds and real estate are two examples) would have done better. Such a display provides an important caveat unavailable in any sort of reportage over a more modest time range.

Of course, we can ask whether even this scale is large enough. Would a different message be conveyed had we started 150 or perhaps 200 years ago? Perhaps, but in preparing such a display we must balance the advantages of the long view with concerns about relevance. Are the forces that yielded the crash of 1929 still operative? Obviously, many of the laws that govern the marketplace are not man-made and hence will always prevail, but some procedural safeguards (e.g., margin rate limitations) are now in place that make some ancient concerns anachronistic.

Thus when we look at, or prepare, a time-based statistical graphic, it is important to ask what is the right time scale, the right context, for the questions of greatest interest. The answer to this question is sometimes complex, but the very act of asking it provides us with some protection against surprises.

15 Sex, Smoking, and Life Insurance: A Graphical View

My mailbox, like those of most American adults, is often filled to overflowing with ads and solicitations of various sorts. Strategically located between my mailbox and my back door is the paper recycling barrel; hence most of this capitalist detritus rarely makes it into the house. But sometimes, either because the topic strikes a chord of interest, or, more likely, because the unrequested commercial message has been hidden inside some more worthy missive, a solicitation makes it inside for more careful examination. I strongly suspected the latter when I found on my mail pile an ad for Jackson National's ten-year level term life insurance.

The ad consisted principally of a postage-paid card offering additional information as well as two large tables of monthly premiums. It was these tables that attracted my attention. The tables provided the monthly premium by age, from 30 to 70, for males and females, smokers and nonsmokers, for $100,000 and $250,000 policies. An extract from these tables is shown in table 15.1. Unlike most American adults, I am fascinated by such tables. I always want to know how it was generated and what it portends. This chapter answers these questions and shows the pathway followed in finding out.

A quick glance at the table provides some obvious information: women pay somewhat less than men, young people far less than old ones, and nonsmokers a lot less than smokers. But most of this is impressionistic and does not provide the sort of deep understanding of the characteristics that govern insurance premiums that could be gleaned from the matrix of numbers provided. How can we learn more?

Because a good graphic display of data is the best way to find what we were not expecting, the first step toward such a deeper understanding is to draw a graph. One such plot, shown in figure 15.1, provides us with a little more understanding, albeit mostly qualitative. We see that premiums skyrocket as we age, but with a real acceleration after age sixty. We also see that the effects of smoking on the monthly premium (and, we infer, on risk of mortality) also increase dramatically after age sixty.

But what does the word *dramatically* mean exactly? As we shall see, in this instance it means exponentially. *Exponentially*, as a word, is different than the word *dramatically* because it has a formal mathematical meaning. Before continuing to look at insurance premiums, it is worth a short digression to explain what *exponentially* means.

The best example of exponential growth comes from the tale about the seventh-century origins of the game of chess. It seems that a very rich Indian rajah was so delighted with the game that he proposed to pay the inventor almost anything he desired for his ingenuity. The inventor asked humbly that for his reward he wanted a grain of rice placed on the first square of the chess board, two grains on the second square, four on the third, and so on for all sixty-four of the chess board's squares. The rajah quickly agreed, thinking that such a price was modest for such a wonderful game. Only later did he discover that, long before the board was completed, the entire annual rice crop of India would be given away. The huge price comes from the repeated doubling. The number of grains to be placed in the kth square is 2^{k-1}. Because the number of the square appears in the exponent, the pattern of growth is known as exponential. Even with the seemingly small base, 2, the result rapidly becomes very large. Thus on the last square the inventor was to receive 2^{63} or 9,223,372,036,854,775,808 grains of rice. The total amount (adding up all of the spaces on the board) is almost exactly twice that amount ($2^{64} - 1$). Such is the power of exponential growth.

If growth is exponential, it can be tamed visually by plotting the data on a logarithmic scale. When we do this (see figure 15.2), an informative result emerges. What was exponentially curved is now straight; actually, the pattern is a sequence of four straight lines,* whose slopes increase at ages 35, 46, and 60. The change in slope is clear at age 35, very subtle at age 46, and profound at age 60. This plot has helped us understand the structure of the relationship between age and life insurance premiums.

* Mathematicians call such a function "piecewise log-linear."

Table 15.1
Jackson National's Ten-Year Level Term Policy Monthly Life Insurance Premiums for $100,000 (in dollars)

	Male		Female	
Age	Nonsmoker	Smoker	Nonsmoker	Smoker
30	12.34	22.34	10.85	17.71
31	12.51	23.23	11.03	17.89
32	12.69	24.21	11.29	18.07
33	12.78	25.19	11.46	18.25
34	13.04	26.26	11.55	18.42
35	13.21	27.41	11.81	18.60
36	13.74	29.01	12.16	19.49
37	14.35	30.71	12.51	20.47
38	14.96	32.57	12.95	21.54
39	15.58	34.53	13.39	22.61
40	16.28	36.67	13.91	23.85
41	17.15	39.25	14.44	25.10
42	17.94	42.10	15.05	26.43
43	18.81	45.12	15.66	27.86
44	19.78	48.51	16.28	29.37
45	20.83	52.07	17.06	30.97
46	22.14	55.18	17.85	32.66
47	23.63	58.56	18.73	34.44
48	25.20	62.21	19.60	36.40
49	27.04	66.04	20.65	38.45
50	28.79	70.13	21.70	40.67
51	30.63	74.67	22.75	43.25
52	32.64	79.57	23.89	46.01
53	34.83	84.73	25.11	49.04
54	37.01	90.34	26.43	52.24
55	39.55	96.30	27.83	55.71
56	42.53	103.06	29.14	58.65
57	45.76	110.27	30.71	61.68
58	49.35	118.01	32.29	64.97
59	53.20	126.38	34.13	68.35
60	57.31	135.37	35.88	72.00
61	63.35	150.77	39.29	79.30
62	70.18	168.03	43.23	87.40
63	77.61	187.35	47.51	96.39
64	85.93	208.97	52.41	106.36
65	95.38	233.18	57.75	117.48
66	106.49	260.59	63.18	130.56
67	119.26	291.30	69.30	145.25
68	133.70	325.74	75.95	161.71
69	149.89	364.37	83.30	180.05
70	168.09	407.62	91.44	200.52

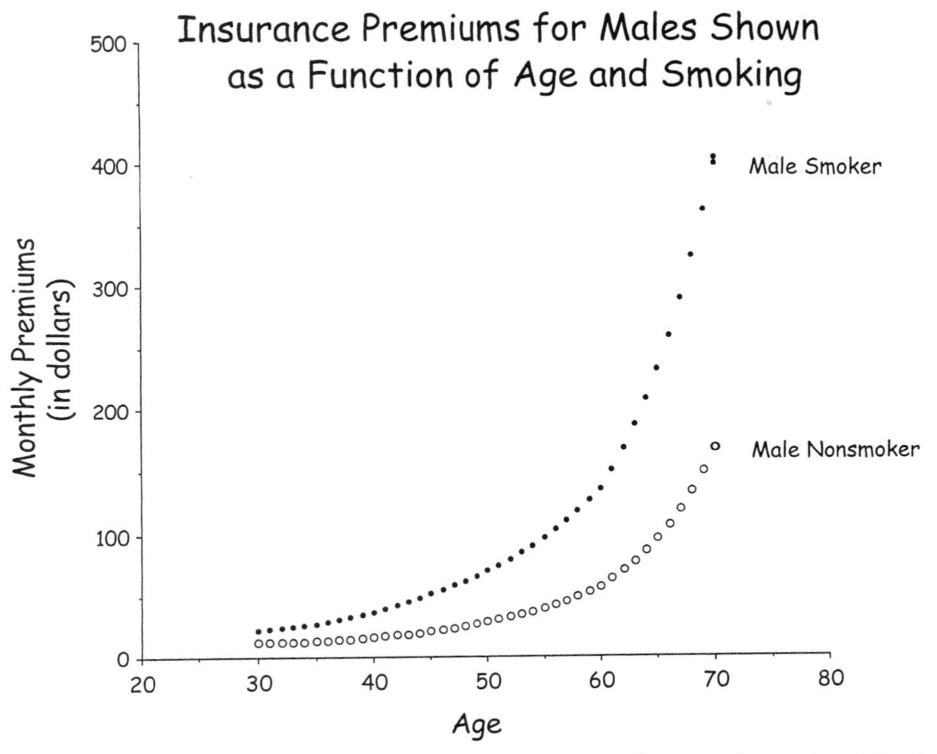

Figure 15.1. Insurance premiums for males shown as a function of age and smoking classification.

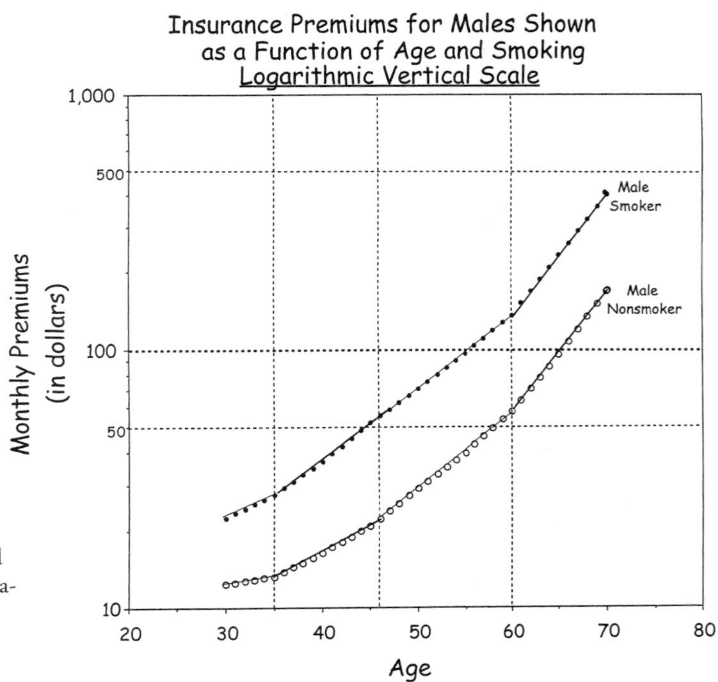

Figure 15.2. Insurance premiums for males shown as a function of age and smoking classification but spaced logarithmically.

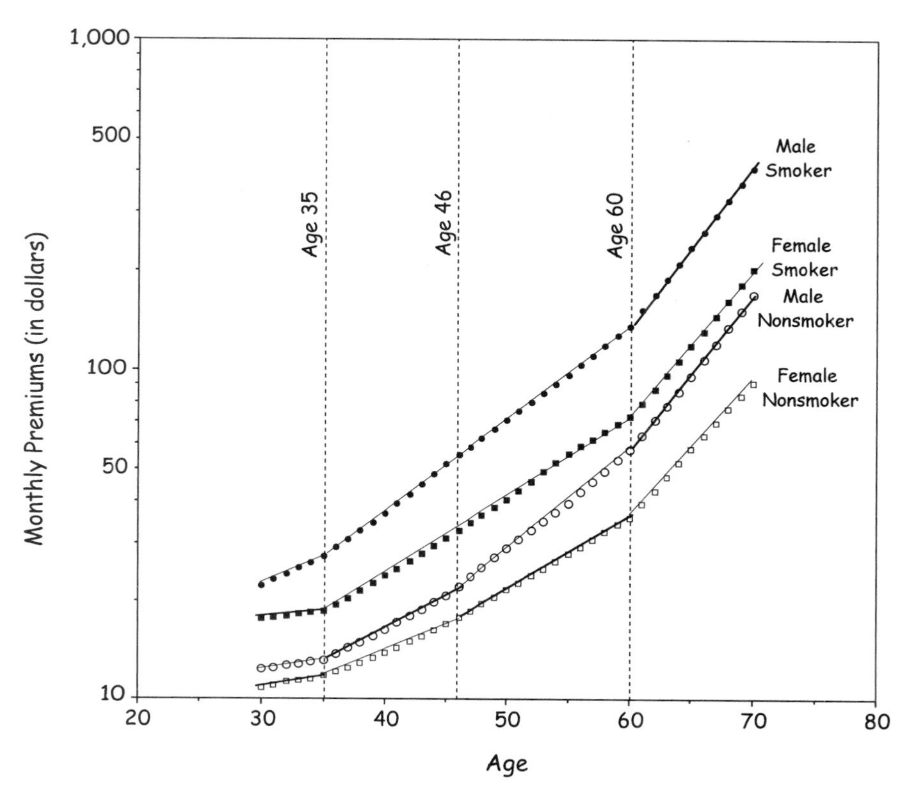

Figure 15.3. Insurance premiums for males and females shown as a function of age and smoking classification and spaced logarithmically.

Now that the secret has been uncovered, we can add women to this plot and see whether the same rules apply. When this is done (see figure 15.3), we find that indeed women have the same general structure of premiums as men, with the break points between the straight segments at the same ages. But the heights of the lines are different, as are the slopes.

We saw earlier the obvious fact that, because of their lower mortality rate, women pay lower insurance premiums than men. But what we now observe is that at all ages smoking causes a *greater* increase in rates (and so, we infer, a greater increase in mortality as well) than does being male.

This brings us to the end of one experiment in analytic detective work, a demonstration of how a graphical tool can help us to decipher the complex world around us. It also shows that the world is not as complex as it might first appear and that the risk estimates that Jackson's National made were evidently calculated from groups of individuals in four age categories.

But even the magic of logarithms does not always tame profound differences. One example of this appeared on July 8, 1997, in the automobile section of the *New York Times.* It listed the names and prices of the forty-seven convertible car models then available. They ranged from the $15,475 Honda del Sol all the way up to the Ferrari F50, priced at a cool $487,000. The range of prices was so broad that the illustrator of the *Times* could not fit all of them on the same graph and so had to resort

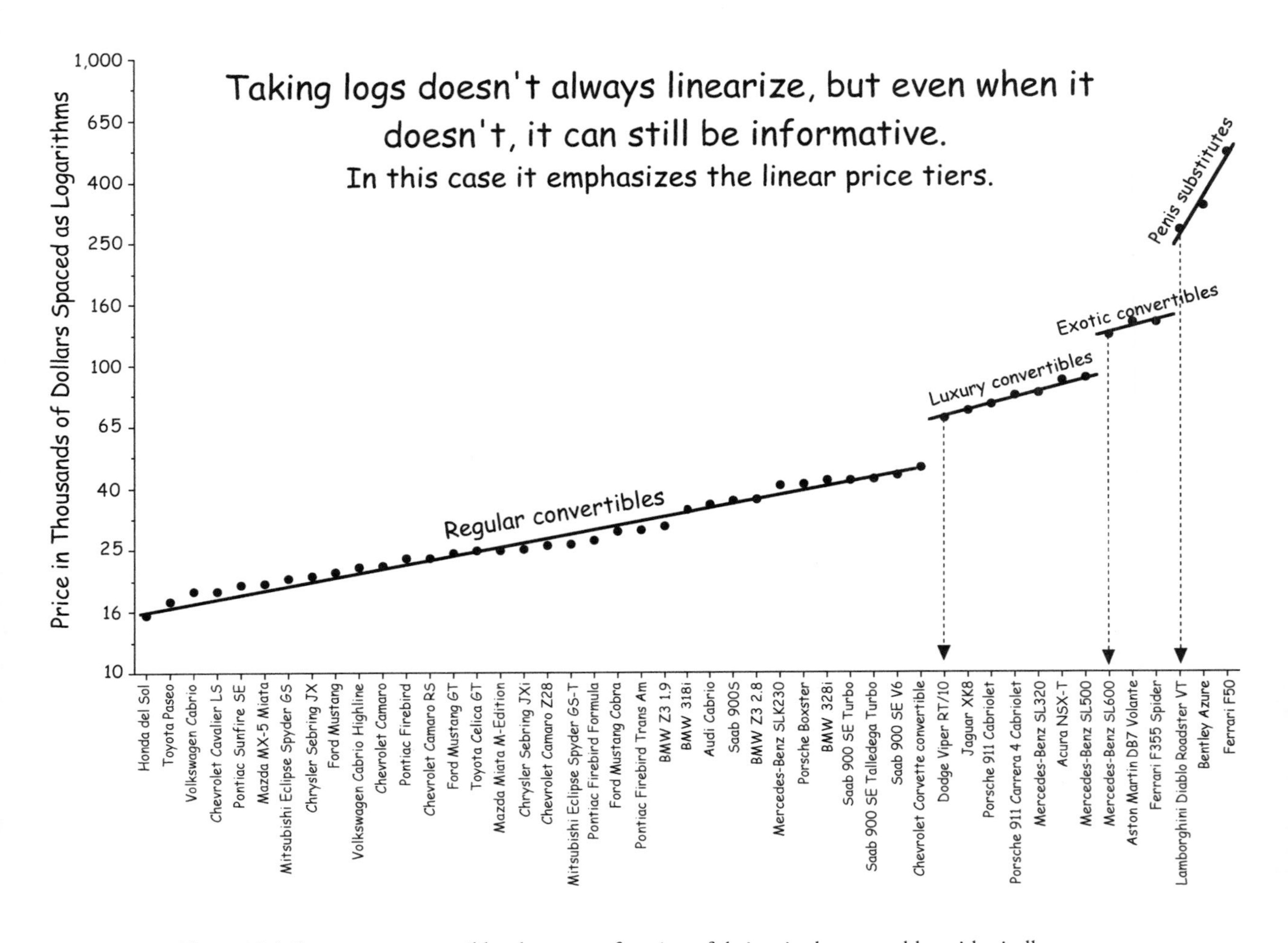

Figure 15.4. Forty-seven convertibles shown as a function of their price but spaced logarithmically.

to merely listing the most expensive ones separately. "Aha," I thought, "you should've tried logs!" Then I did, figuring that they would linearize the plot.* They did, but not all cars fell on the same straight line (see figure 15.4). The thirty-four least expensive cars fell on one straight line, then there appeared to be a jump for the next seven cars, then another jump for three exotic European cars, and then on another line with a much steeper slope were the three most expensive cars.

Thus we have discovered that the cars at the upper reaches of the domain have prices that exceed what a single exponential increase would dictate. Would some other way of thinking about convertible prices provide a clear functional relationship between rank order and price?

* The expectation that the log of prices of convertibles would be linear when plotted against the rank order derives from Zipf's Law (Zipf 1949; Wainer 2000a). Because it deals with frequencies, Zipf's Law does not appear to be strictly applicable here, yet it holds so often that we are surprised when it does not.

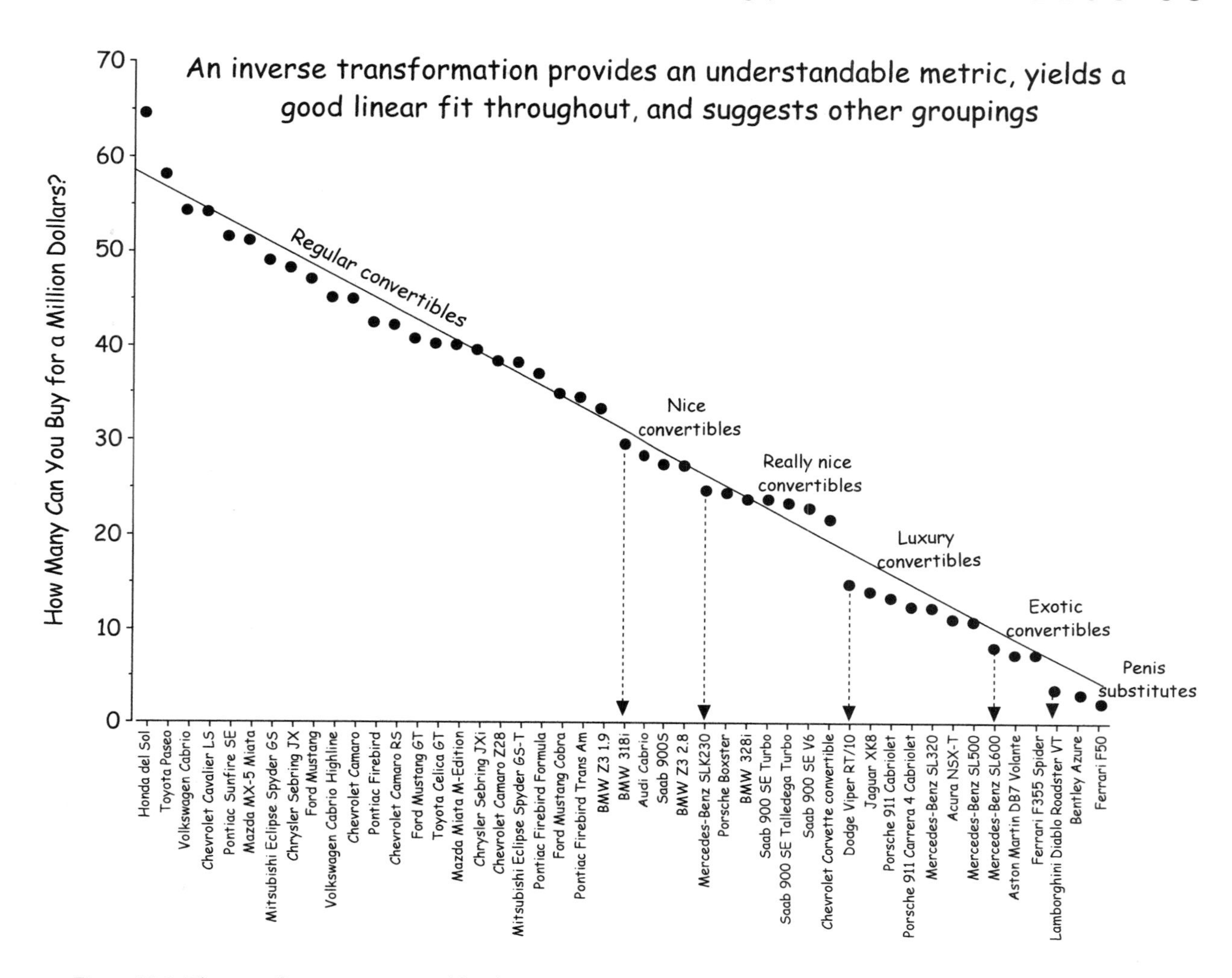

Figure 15.5. The same forty-seven convertibles shown as a function of their inverse price (how many of that model could be bought for a million dollars).

Taking the next step (inverses) in this ladder of transformations does the trick.* If we make a plot showing how many of each car we could buy for one million dollars (about sixty-five Honda del Sols but only two Ferrari F50s), a simple linear relationship emerges (see figure 15.5). By making this transformation, we not only gain a linear relationship between rank and a function of price, but we also have a metric that is easier to think about and to communicate to others. We also discover that two other groups of cars separate themselves from the pack: "nice convertibles" such as the BMW Z3, and "really nice convertibles" such as the

* Traditional statistical summaries, like the mean, work best when the distribution of data being summarized is symmetric. Anyone who has gone to a restaurant and eaten a cheeseburger when everyone else had the lobster understands that dividing the bill equally (averaging) does not yield a satisfying result. One approach when data are skewed is to transform them to symmetry. For mild skewness, a square root often works; when it is more extreme, a logarithm; when more extreme still, an inverse. These steps in progression are often called "Tukey's ladder of transformations." See Tukey 1977.

Porsche Boxster. Their separation is clearer in the inverse metric than it was in the log metric (although, now that we know to look for it, we can also find it in figure 15.4).

The Chicago statistician David Wallace used to include the dictum "The second step in analyzing data is to take their antilogs" as part of the introductory lecture that he gave in most of the courses he taught. I have been amazed, in the almost thirty years since I first heard him say this, at the wisdom of this advice. In fact, I have found that it is an unusual data set indeed that yields its secrets more readily when it is left untransformed.

16 There They Go Again!

I am addicted to the *New York Times* and feel restive if I miss it for even a day; after two days I become seriously out of sorts; longer periods of abstinence are almost intolerable. When the *Times* makes an error, I feel the same sort of disappointment that one would evince at the discovery of a flaw in any loved one. In the past I have tried gently to point out graphical flaws in *Times* reportage,[1] in the hope that the errors would be corrected. It is hard to measure the effect that my efforts have had.

In general, I have been happier with the statistical graphics in the *Times* than in most other print media, but when three pretty serious problems, two graphical and one statistical, showed up in the same display, it seemed clear that some remedial action ought to be taken.

Problem 1: Nonlinear Time

Time usually moves in an inexorable linear fashion (at least when we are all moving at about the same speed). Thus, when a graph distorts time, the image of the phenomenon depicted is often seriously affected. Figure 16.1 is a plot, closely resembling one that originally appeared in the *Washington Post* (January 11, 1979),* that shows the linear growth of physicians' incomes over nearly four decades starting in 1939. It takes a sharp eye to spot the nonlinear time pattern. When the data are replotted in a more consistent manner (figure 16.2), we see that physicians' incomes have increased exponentially.

Whatever is the affliction of *Post* graphics that causes time distortions, it seems to have migrated north and infected my beloved *Times*. On June 22, 1997, the *Times* published a collection of six data displays (five graphs and a table, reproduced as figure 16.3); three of the graphs contain a flaw.

The very first one shows two lines depicting the change in life expectancies for men and women over a fifty-five-year period beginning in 1940. The message I drew was that increases in life expectancy had leveled off and that the gap between men and women seems pretty constant. But a closer look shows that they did it again! The time scale has changed with a fivefold expansion at 1970. When this is redrafted (figure 16.4) with the time axis held to a constant scale, we discover that life expectancies continue to increase linearly and that the gap between men and women shows hints that it is widening.

I am grateful to Richard Thayer for directing my attention to the material in figure 16.3; the chapter title paraphrases Ronald Reagan's line from a 1980 debate with President Jimmy Carter.

* The original plot is not reproduced here because the *Washington Post* was unwilling to sell the right to reprint this graph for criticism.

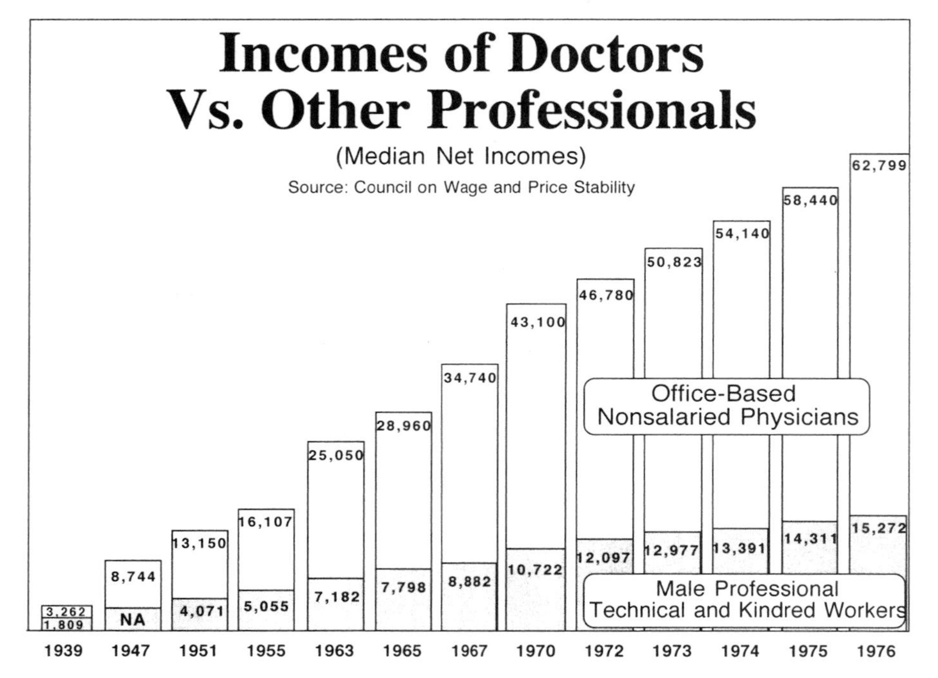

Figure 16.1. Changing the scale in mid-axis to make exponential growth seem linear (after a plot in the *Washington Post*, January 11, 1979, in an article titled "Pay Practices of Doctors on Examining Table" by Victor Cohn and Peter Milius).

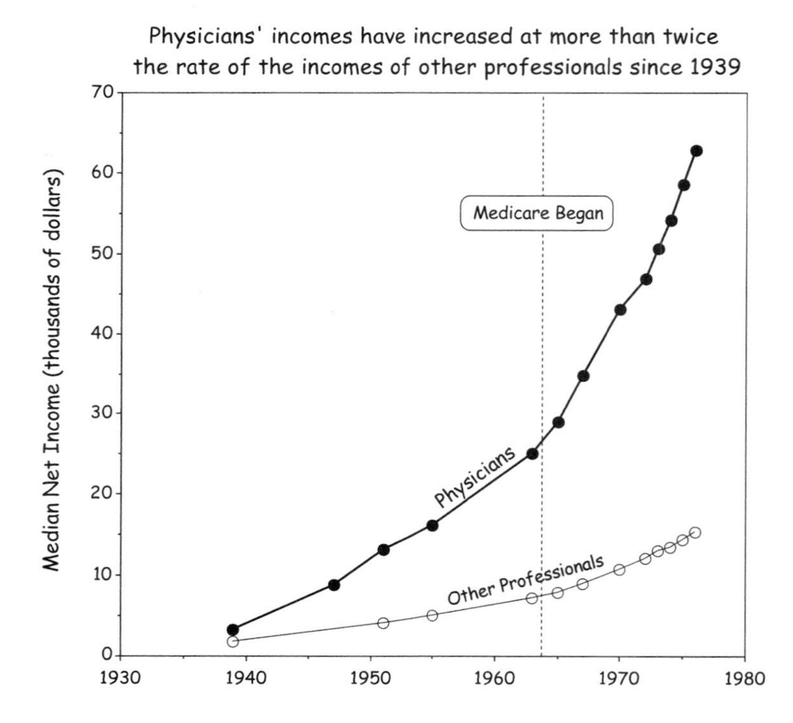

Figure 16.2. Data from figure 16.1 redone with time on a linear scale (from Wainer 1980a).

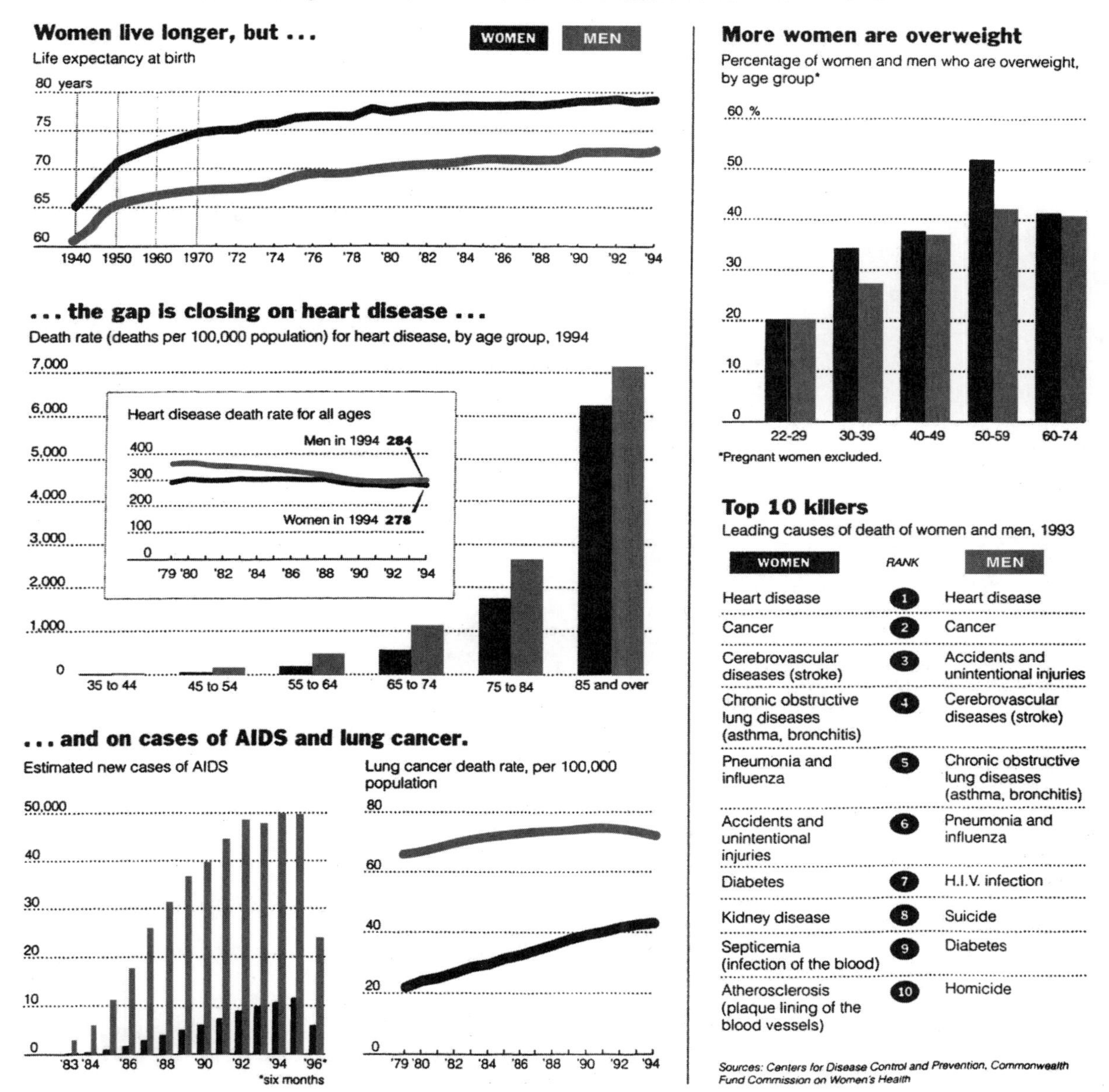

Figure 16.3. A data display from the *New York Times* (June 22, 1997, page 24 of the Women's Health section) that contains five graphs. Three of those graphs have serious flaws.

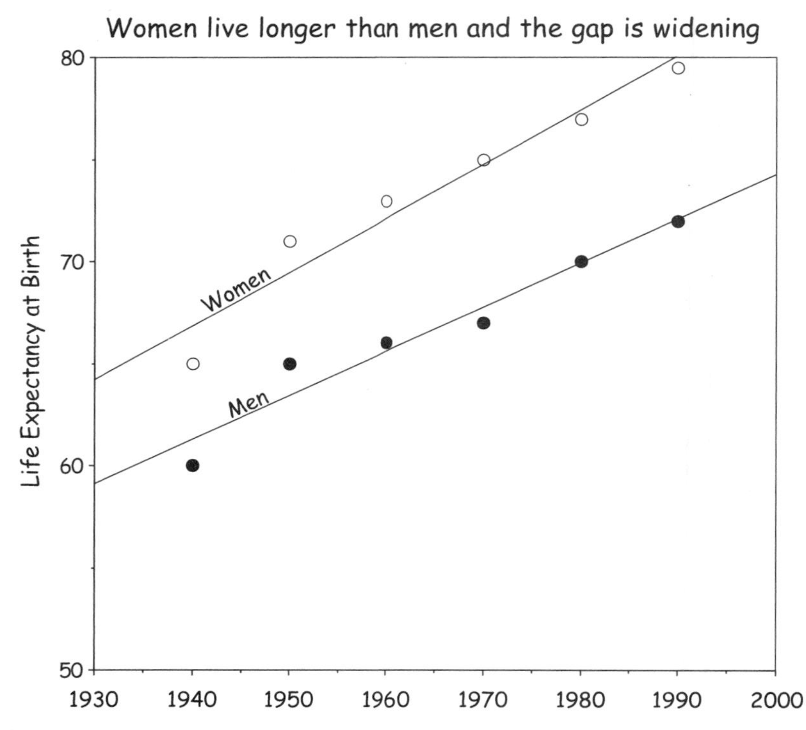

Figure 16.4. The data from the upper left-hand panel of figure 16.3 redrawn with time on a linear scale show continued gains in life expectancy and a slight widening of the gap between men and women.

Problem 2: An Inconsistent Metaphor

The bottom left-hand panel of figure 16.3 is reminiscent of an earlier *Times* plot (figure 16.5), which accompanied a story that proclaimed that the recent trend in airline fare cuts has had a deleterious effect on the fees paid to travel agents. In fact, the headline, above both the story and the graph, reads "Air Travel Boom Makes Agents Fume: Fare Cuts Lower Fees." Yet a closer look at the labels shows that the third bar—the bar that supports the theme of the story—is only for the "first half [of] '78." This bar omits the heavy travel period at the end of the summer as well as travel around the Thanksgiving and Christmas holidays. But most grievously, it omits the second half of 1978. Figure 16.6 shows what transpired once the second half's commission payments were included. Contrary to the story, it was a banner year for travel agents.

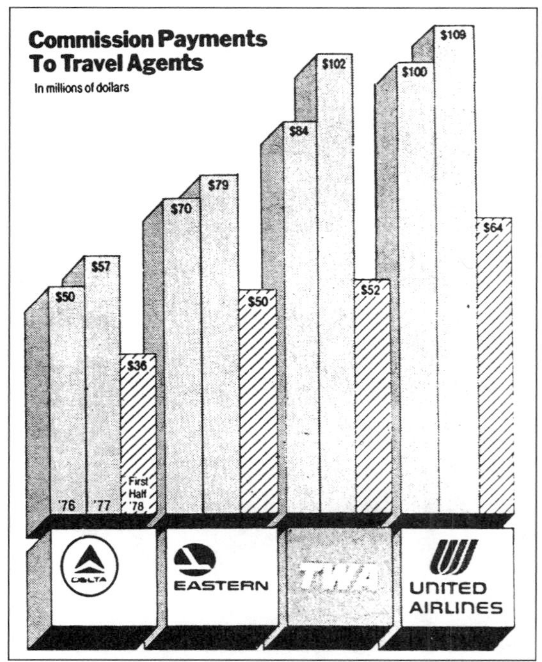

Figure 16.5. Mixing a changed metaphor with a tiny label reverses the meaning of the data (from the *New York Times*, August 1978).

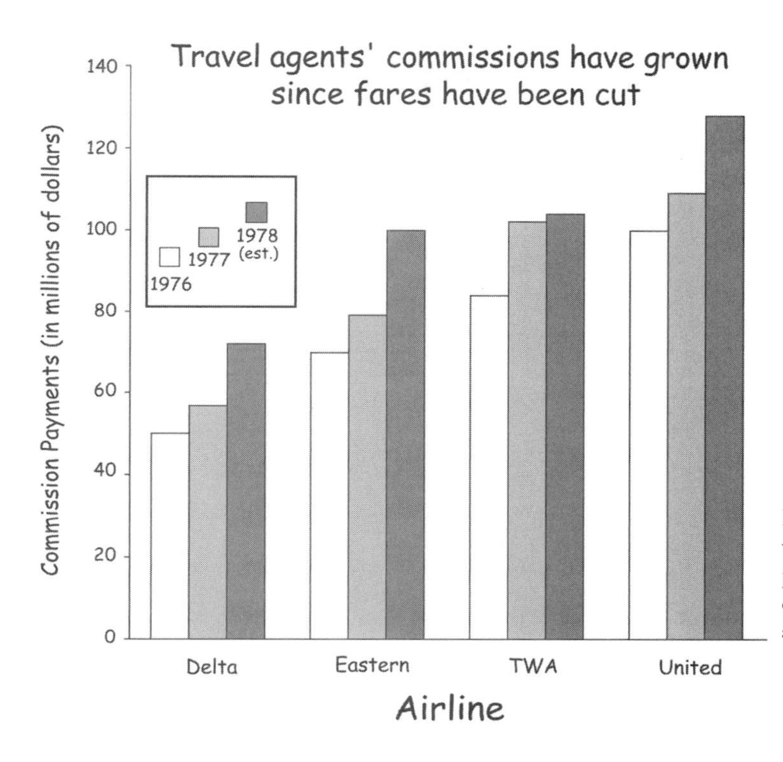

Figure 16.6. Figure 16.5 redrawn with 1978 data placed on a comparable basis shows that the fare cuts have been a boon to travel agents (from Wainer 1997b).

But what about the incidence of new cases of AIDS? The bottom left-hand panel of figure 16.3 suggests that there was a peak in new AIDS cases in 1994 and 1995 and that there was a rapid decline in 1996. But now there is an ominous asterisk with an associated footnote that tells us that this is just for six months. If we double the amounts shown, roughly annualizing the estimator for 1996, we find (figure 16.7) that new AIDS cases appeared with nearly constant frequency during the five years 1991–1996 for both men and women.

Problem 3: Simpson's Paradox and Heart Disease

The middle panel in figure 16.3 is an interesting demonstration of statistical naïveté. The age-specific death rates show quite clearly that at all ages, men die more often from heart disease than do women. Yet the inset, which averages across ages, shows that the men's death rate from heart disease has dropped. This leads one to suspect that some sort of Simpson's Paradox–like anomaly is taking place (see chapter 10), but the headline writer seems to have missed it. Perhaps the age-specific death rates in earlier years would support the conclusion, but the image conveyed—that the sex difference in death rates from heart disease has shrunk to almost nothing—seems clearly wrong.

In his *Life of Reason II*, George Santayana paraphrased Thucydides (in his *History of the Peloponnesian War*, book 1, section 1) when he wrote, "those who cannot remember the past are condemned to repeat it." It was my goal here to help my beloved *Times* remember the past so that it does not repeat it quite so often—at least not so often on the same page. I really don't much care what the *Washington Post* does.

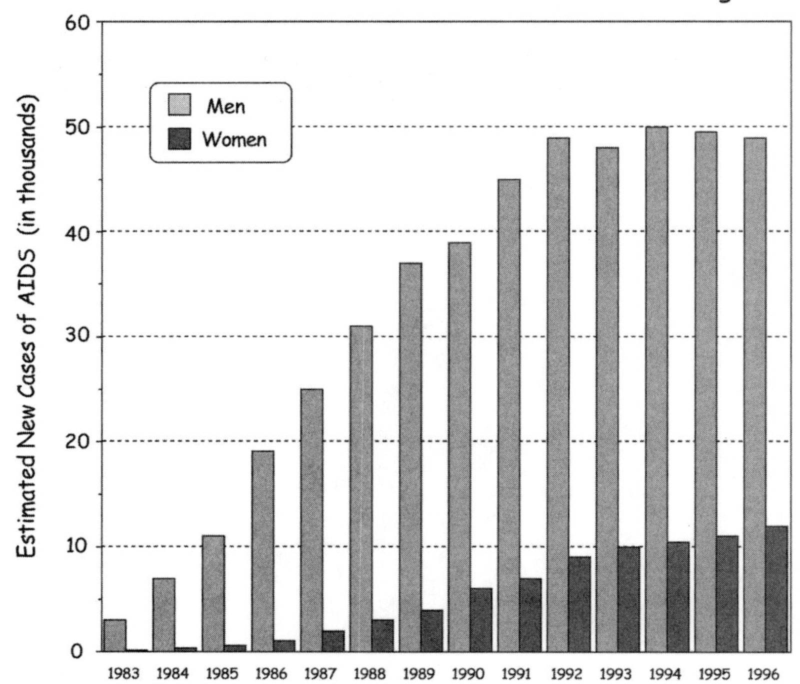

Figure 16.7. The data from the lower left-hand panel of figure 16.3 redrawn with 1996 data placed on a comparable basis show that the number of new AIDS cases remained steady from 1992 to 1996.

17 Sex and Sports: How Quickly Are Women Gaining?

Over the past century, women's participation in competitive sports has increased markedly. At the same time the disparities in performance between men and women seem to have shrunk substantially. One cannot help but suspect that the former is a principal cause of the latter.* We can see this effect clearly in data from the Boston Marathon, which has been run over essentially the same course for more than a century. A plot of winning times is shown in figure 17.1.

Such evocative plots have led the unwary to

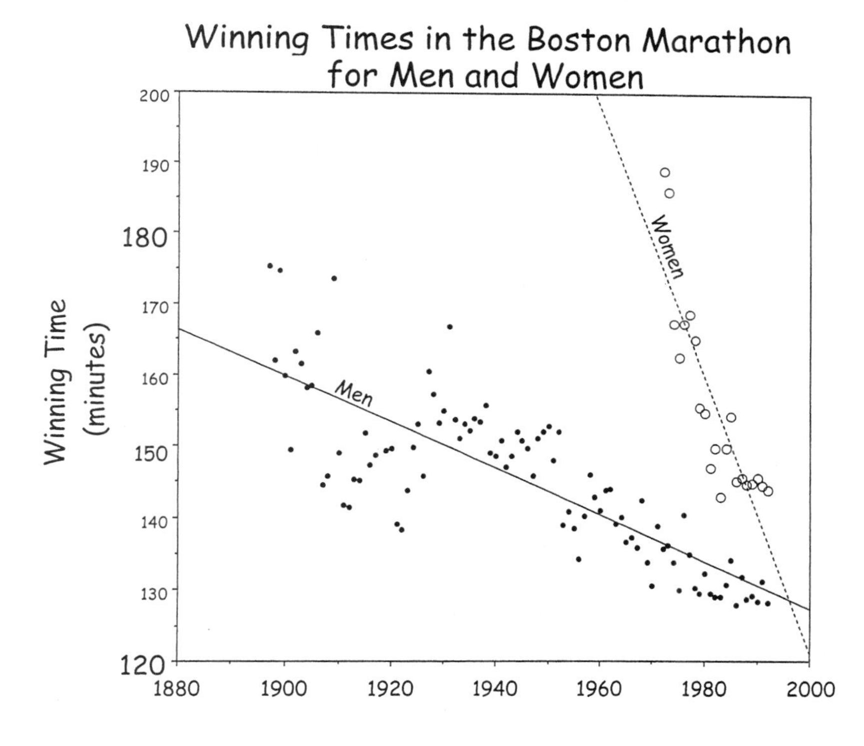

Figure 17.1. All of the winning times in minutes for men and women in the Boston Marathon over the last century. A fitted straight line to each data string summarizes and predicts.

This chapter is abstracted from a longer paper that I published with Samuel Palmer and Catherine Njue (Wainer, Palmer, and Njue 2000). I am delighted that they have both given their permission to reproduce a subset of our work here.

* Scott Berry (2002) shows how a simple model of population size accounts for a very large proportion of the athletic gains over the past century in sports. This provides strong support for the view that participation rate is key to understanding improvements in athletic performance.

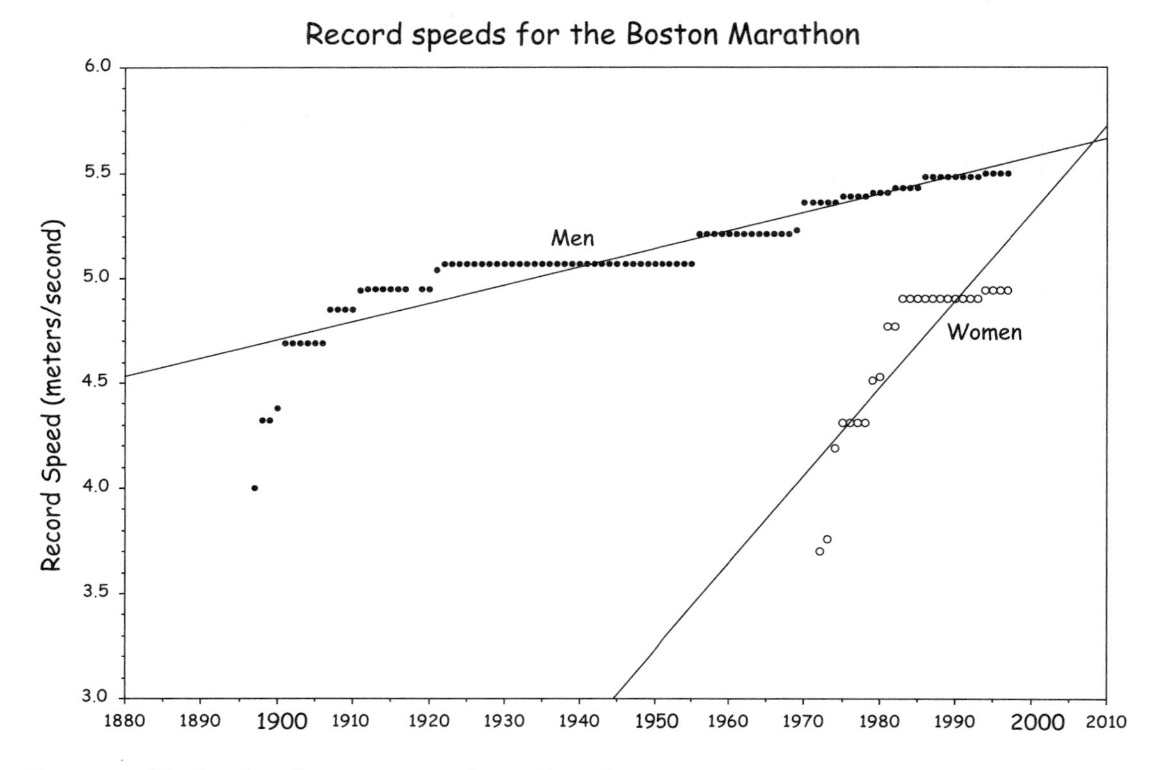

Figure 17.2. The data from figure 17.1 inverted to yield speeds in meters per second. The fitted lines suggest the same inferences as those fitted to winning times.

predict that women runners will soon be overtaking their male counterparts and, indeed, would break the two-hour barrier in the marathon sometime shortly after the new millennium began.

For a variety of reasons, primarily to ease comparisons with other events, it is useful to transform these winning times to speeds (in meters per second) and to consider only record times.* If we do this, the data in figure 17.1 can be represented as in figure 17.2. Once again a line fit to the two data sets suggests women overtaking men relatively shortly at speeds of roughly 5.5 meters/second.

Most experts in track and field do not find this extrapolation credible.

In figure 17.3 we can see that the curve that describes the improvements in men's performance in the very early years of the race, when participation among men was limited, is quite similar to the curve that describes the women's performance since 1972, when they were first allowed to compete. As women's participation increased, so too did the speed

* Athletic performance is yet another example in which it is advantageous to transform into a different metric. Just as the price of the convertibles in chapter 15 was more profitably examined after an inverse transformation, so too it is for running times. It is interesting to note that the North American metric for judging automobile fuel efficiency (miles per gallon) is not the same one used in continental Europe, which favors its inverse (number of liters per hundred kilometers). In this instance the European practice is more useful. Consider the problem "I have 300 miles to travel and I get 18 miles to the gallon. How much fuel will I need?" and compare its difficulty of solution to ". . . and I use 5.5 gallons per 100 miles." The gallons should be in the numerator because you pay for them—the miles come free.

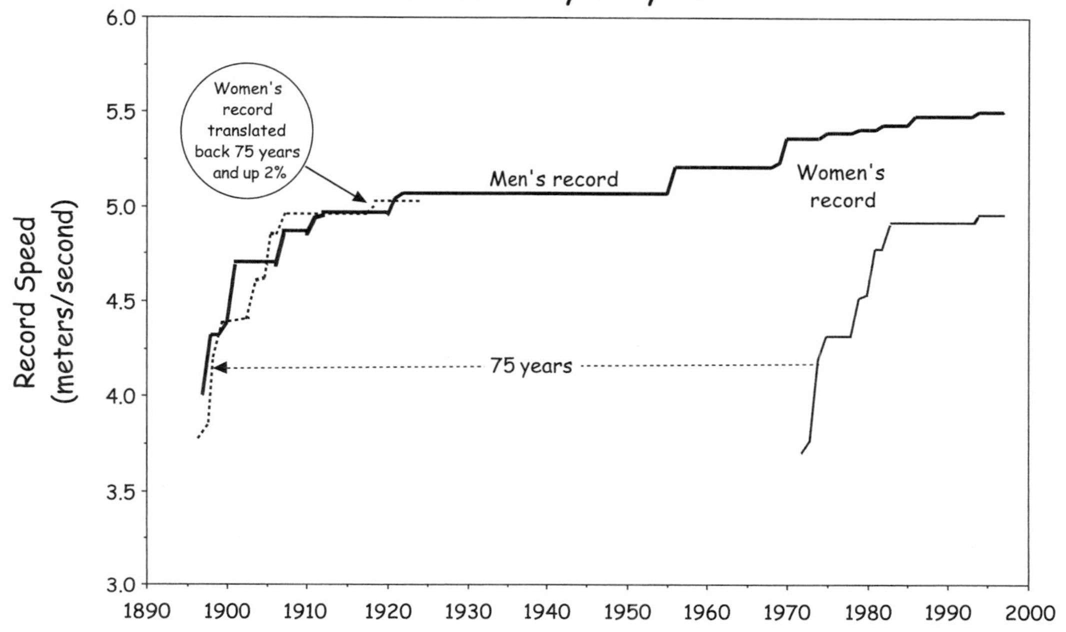

Figure 17.3. The curve of women's winning speeds in the Boston Marathon, when shifted upward by 2 percent, matches the men's curve seventy-five years earlier.

necessary to win. How can we characterize this simply? Suppose we duplicate the data points associated with the women's record speed and move them to where they roughly coincide with the men's records (figure 17.3). This suggests a different approach to representing the difference between men's and women's performances.

This shift of the women's data, what mathematicians might call "a translation to registration," essentially provides an answer to the two-part question "how many years ago was the men's record 2 percent faster than the women's record is today?" The two parts of this question are: "how many years ago," which is a translation of the curve to the left, and "2 percent faster," which is a translation upward. The key assumption here is that the improvement in performance for both men and women follows essentially the same curve, and so to make accurate comparisons we must translate one curve to overlay the other.

When we plot the difference in time between men's and women's records (in figure 17.4), we find that it is remarkably flat, suggesting that between sixty and eighty years ago, men were running the Boston Marathon about 2 percent faster than women are now and that the improvements that women have shown over the past twenty years were mirrored by improvements that men made more than a half-century earlier. This result more closely matches experts' opinion in their estimation of when women will overtake the current men's record in the Boston Marathon.

Now that we have established the plausibility

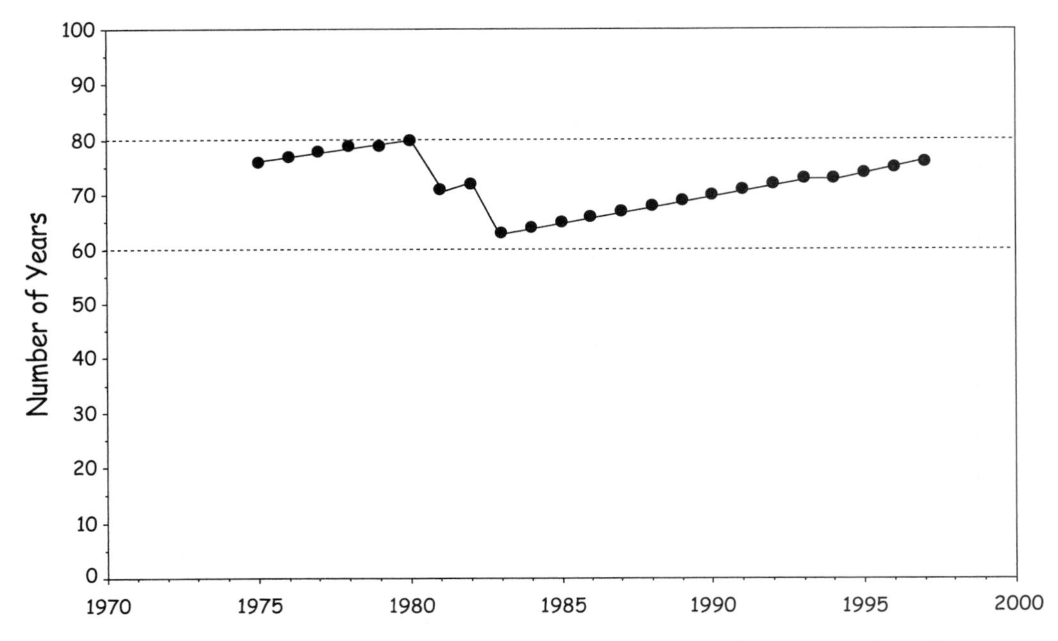

Figure 17.4. A plot of the number of years that have passed since the men's record was the same as the women's record (made 2 percent faster). The suggestion of a seventy-five-year shift is borne out.

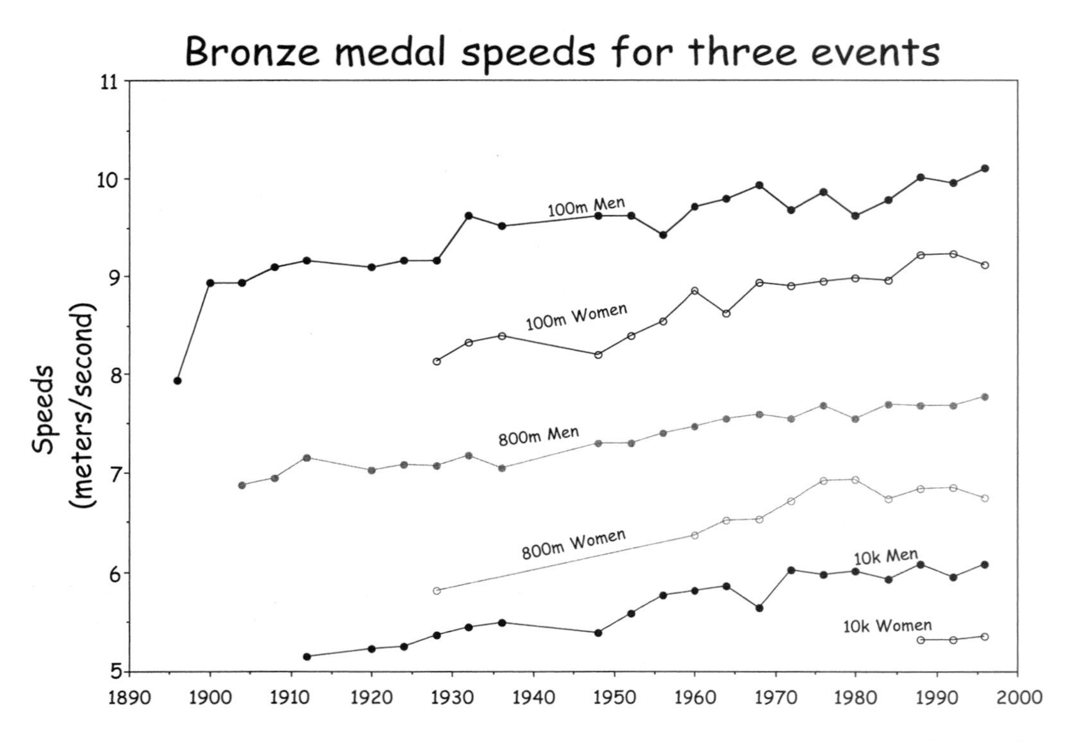

Figure 17.5. The bronze medal speeds for three Olympic track events show very similar gains for both men and women. This is true for a sprint, a middle-distance, and a long-distance event.

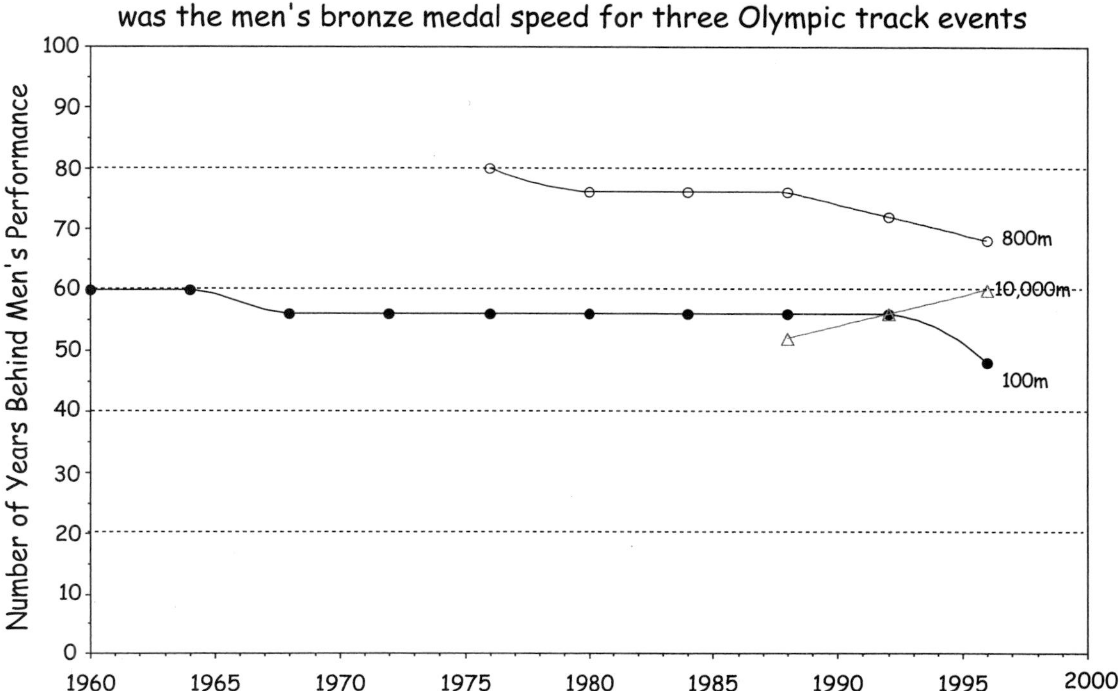

Figure 17.6. Using the same idea of looking at time shifts, we find that women's performances lag those of men by fifty to seventy years, but the lag is least in the sprint race.

of this measure as a way of comparing men's and women's performances, let us briefly consider women's progress in three other track events; the 100 meter, 800 meter, and 10,000 meter runs. To examine these, we shall use the bronze medal speeds from the Olympic Games. We use the bronze rather than the winning (gold medal) speeds for statistical stability.* The raw results are shown in figure 17.5. We can see that the improvements in men and women in all three of these events (as viewed by the fitted trend lines) are roughly parallel. That is, men and women Olympians seem to be improving at about the same rate in all three events examined. And this seems to have been true for a long time.

We next do the same sort of shifting with these curves as we did earlier with the Boston Marathon and plot the amount of shift. This graph is shown in figure 17.6. These results are consistent with what we saw earlier from the Boston Marathon, with one important and interesting variation. Specifically, the women's speeds for the 800-meter run are about what men's speeds were seventy to eighty years ago, but for the 100-meter run,

* The bronze medal time is the median of the top five finishers. The winning time can sometimes not be representative, whereas taking a median of the top five yields a more statistically stable measure.

women are lagging by only fifty to sixty years. This runs counter to the common conception of women's performance in longer races being closer to that of men's. By this measure at least, this is not true. It seems prudent to suspend interpretation of the 10,000 meter race until more data points are available.

Returning to the title question, "How quickly are women gaining?" it appears that we must respond, "They aren't." Indeed, according to this measure, the gains women have shown over the past decade or two in track performances have served only to prevent them from falling farther behind.

18 Clear Thinking Made Visible: Redesigning Score Reports for Students

> Good information design is clear thinking made visible, while bad design is stupidity in action.
>
> —Edward Tufte, "Visual Explanations"

Edward Albee, in his one-act play *Zoo Story*, points out that sometimes you need to go far out of your way to come back a short distance correctly. While what follows may seem to meander, please stay with me.

In 1980, when I began working at the Educational Testing Service, my boss, Paul Holland, gave me some important advice: "An effective memo should have at most one point." Over time, whenever I have violated this advice I have never been happy with the result. I think the lesson explicit in Holland's Rule can be broadened.

I have had a thirty-seven-year-long relationship with Princeton University, first as a student, next as an alumnus and sometime faculty member, and now, my newest role, as the parent of a student. Throughout this time, in all of these roles, I have always been impressed with the sagacity of both the form and the content of the communications the university has had with its various constituencies. The university seems to craft both the form and the content of its messages carefully for their specific purposes. But even though my admiration was considerable, it grew still further when I read the response that my son received to his application to the university. An accurate facsimile of that missive is shown as exhibit 18.1.*

The elegant spareness of this letter demonstrates a deep understanding of the desires and needs of its audience. By clearly saying everything necessary, but no more, it provides a fine model for other sorts of communications.

The university's letter provided a contrast to another college entrance communication that we had received a year or so earlier. An envelope arrived at our home, addressed to my son, which we hoped contained his college admissions test scores. The envelope lay unopened on the kitchen table while the addressee's anxious parents awaited his return home from swim practice. When he finally arrived and opened it, he began reading silently. What felt like several minutes elapsed before we finally asked, "Well? How did you do?" He replied, "I don't know. I'm still looking for my scores."

To understand my son's predicament, consider

* I should point out that although this letter was the principal focus of the communication, it came in a large envelope with many other things, among them a "free" Princeton decal suitable for a car. When my son asked for permission to attach it to the window of the family car, my wife complained, "Sam, you're trying to turn me into a suburban cliche." He replied, "But mom, you already have the Volvo station wagon." The Volvo now proudly sports this sign of our collegiate allegiance.

PRINCETON UNIVERSITY **Admission Office**
MAILING ADDRESS: Box 430, Princeton, New Jersey 08544-0430
OFFICE: 110 West College
TELEPHONE: 609-258-3060 FACSIMILE: 609-258-6743

Fred A. Hargadon
Dean of Admission
(on leave until January 2001)

Steve LeMenager
Acting Dean of Admission

December, 2000

Dear Sam,

Yes!

We are happy to offer you admission to Princeton University and are delighted to welcome you as a member of the class of 2005.

Sincerely,

Steve LeMenager
Acting Dean of Admissions

Exhibit 18.1. Princeton University's acceptance letter.

the generic score report shown as exhibit 18.2. There is a great deal of information in it, but the differential visual impact of its various components does not match the differential interest that the reader has in each aspect.

How should such a report be designed? As in all design situations, it must begin with an understanding of purpose. What are the questions that a person receiving such a score report wants answered? I suspect that there are four basic questions that require an answer. They are, in roughly their order of importance to the examinee:

1. What is my score? This might be a single number or a set of numbers.
2. How do I compare to others? A fact without context is of little value. A single number tells us nothing without the ancillary knowledge about how everyone else did. Even so-called criterion-referenced tests have latent in them the performance of a reference population. Thus, a four-minute miler is applauded even if he ran alone on the track, but the "objective criterion" of four minutes gets its meaning from our knowledge of how many have tried to do it and how few have succeeded.

SAT Program

Your Scores

Test Date: January 2001

SAT I Reasoning Test	Score	Score Range	Percentiles College-bound Seniors	
			National	State
Verbal	710	670-750	97	97
Math	730	690-760	98	98

Student Score Report

Report Date 9/01/01

Seq# 000000012
Jane Doe
12 Main Street
Hometown, NJ 12345

What does your score range mean?
No single numerical score can exatly represent your reasoning skills. If you had taken different editions of the test within a short period of time your performance would probably vary somewhat on the 200 to 800 point scale.

How do you compare with college bound seniors?
Percentiles indicate what percentage of test takers earned a score lower than yours. The national percentile for your verbal score of 710 is 97, indicating that you did better than 97% of the national group of college-bound seniors. The national percentile for your math score of 730 is 98, indicating you did better than 98% of the national group of college-bound seniors.

Did you do better in verbal or math?
Your score indicate that you performed similarly on the math test and the verbal test.

What's the average verbal or math score?
For college-bound seniors in the class of 1999, the average verbal score was 505 and the average math score was 511.

Will your scores change if you take the test again?
If you take the test again, especially if you study between now and then, your scores may go up.

Among students with verbal scores of 710, 63% score lower on a second testing and 37% score the same or higher. On average, a person with a verbal score 710 loses 21 points.

Among students with math scores of 730, 65% score lower on a second testing and 35% score the same or higher. On average, a person with a math score 730 loses 18 points.

Exhibit 18.2. A sample of a score report for the SAT that is sent to more than one million students annually.

3. How stable is my score? If you stand on a bathroom scale and it reads 100kg, how much will it change if you get off and then get on again?

This ordering represents the typical priority.* Thus the visual emphasis given to each question should reflect this prioritization. In addition, there is a fourth question, strongly related to the first three, whose answer is too often left implicit:

4. What does my score mean? Obviously, this is a validity question, and its answer depends on the score level, how that score compares with others, and how stable the score is. The precise form of this question varies with who asks it. But the answer is almost always a probability statement. For the individual, the question might specialize to "How likely is it that I will get into Princeton?"

The report shown as exhibit 18.2 answers the first question—710 in verbal skills and 730 in mathematical ones—although there is no visual emphasis to this response. These scores are hidden away with the other numbers.

The second question is answered, but giving national and state norms does not provide enough specificity to be of much help.

The third question is answered quite thoroughly, both through the inclusion of what I assume to be plus and minus one standard error bounds and the textual details about the likelihood of gains upon retesting.

* Evidence supporting this prioritization was gathered in a survey of educational policy makers by two University of Massachusetts researchers (Hambleton and Slater 1996). This result was confirmed in subsequent experiments on this same class of test users (Wainer, Hambleton, and Meara 1999).

SAT Program

Test Date: January 2001

Student Score Report
Report Date 9/01/01

Jane Doe
12 Main Street
Hometown, NJ 12345

Scores for Jane Doe					
		Percentiles among college-bound seniors			
			Among the applicants to the colleges you applied to		
SAT I: Reasoning Test	Score	National	Princeton	Cornell	Rutgers
Verbal	**710**	97	47	75	97
Math	**730**	98	57	77	98

How do you compare with college bound seniors?
Percentiles indicate what percentage of test takers earned a score lower than yours. The national percentile for your verbal score of 710 is 97, indicating that you did better than 97% of the national group of college-bound seniors. The national percentile for your math score of 730 is 98, indicating you did better than 98% of the national group of college-bound seniors.

What's the average verbal or math score?
For college-bound seniors in the class of 1999, the average verbal score was 505 and the average math score was 511.

How do you compare with the other seniors who have applied to the same colleges you have?
Your test performance ranks you at:
Princeton about average among all their applicants,
Cornell among the top quarter of their applicants,
Rutgers among the top 2% of their applicants.

Princeton accepts about 10% of all applicants
Cornell accepts about 25% of all applicants
Rutgers accepts about 50% of all applicants

Exhibit 18.3. A suggested revision of the SAT score report that emphasizes the answers to questions that are likely to be of immediate importance to its recipients.

The fourth question is not addressed explicitly.

An effective redesign ought to consider the recipient's priorities. One such redesign is shown as exhibit 18.3.

It makes the actual scores big and obvious (printing them in red would be nice, although using colors is oftentimes too expensive to be practical). Second, since examinees often indicate which colleges should receive their scores, it is straightforward to provide the score distributions for each of the schools that are receiving this person's scores, or at least this individual's place in those score distributions. If each college shared the score distribution of those who were accepted the previous year, a still more informative report could be prepared. Lacking this, we can still provide the overall acceptance rate for each school, which gives a reasonable answer to the all-important fourth question.

This redesign has essentially ignored any mention of score stability. Why? I certainly believe that the stability of scores (their *reliability*, to use test theory jargon) is important, but in most cases neither the applicant nor the schools receiving the scores care a lot about such information (other than the understanding that such scores are stable enough for their intended use). College admission tests are, to a very large extent, a contest. The focus is on who did best on a particular day. Olympic gold medals are not given out to the best athlete determined by averaging a large number of performances over a year or two.* No, it is based

* Olympic figure skating is a clear exception to this, in that skating judges seem to be including each skater's performance in prior competitions and perhaps even in practice in their ratings.

on who won that day. And, in the same way, variability in scores is, at best, a secondary issue and as such can be relegated to the second page of the report. Or maybe the third.

Similar score reports are sent to institutions, and although the goals for such reports are different, the same rules still apply. I strongly suspect that it is only by following these rules that score reports can approach the elegance and efficiency of communication that is the hallmark of Princeton's marvelous letter.

Graphical Displays in the Twenty-first Century

Who of us would not be glad to lift the veil behind which the future lies hidden; to cast a glance at the next advances of our science and at the secrets of its development during future centuries? What particular goals will there be toward which the leading . . . spirits of coming generations will strive? What new methods and new facts in the wide and rich field of [scientific] thought will the new centuries disclose?

History teaches the continuity of the development of science. We know that every age has its own problems, which the following age either solves or casts aside as profitless and replaces by new ones. If we would obtain an idea of the probable development of . . . knowledge in the immediate future, we must let the unsettled questions pass before our minds and look over the problems which the science of today sets and whose solution we expect from the future. To such a review of problems the present day, lying at the meeting of the centuries, seems to me well adapted. For the close of a great epoch not only invites us to look back into the past but also directs our thoughts to the unknown future. (Hilbert 1902, p. 437)

So begins what was perhaps the most influential speech ever made in mathematics. It was David Hilbert's (1862–1943) presentation to the International Congress of Mathematicians in Paris in July 1900. In it he described what he felt were the twenty-three most important unsolved problems in mathematics. His choice was wise and the formidable task of solving these problems occupies mathematicians to this day. Although most of these problems have now been satisfactorily resolved,* there remains much to be done.

Hilbert's strategy was brilliantly successful and so, as we approached the millennium, it seemed sensible to me to emulate him as best I could and examine the problems that currently confront data display. Then, I might do as Hilbert did and present these problems to the field and wait for answers. But I am twenty years older than Hilbert was when he proposed his problems, and so must be forgiven my lack of patience in such a strategy. Happily, I had at my disposal a resource for solving these problems that was unavailable to Hilbert—I could ask John Tukey for the answers.

* Eighteen of Hilbert's problems have been solved; two (problems 14 and 20) proved to be false. To the best of my knowledge, only three of Hilbert's problems truly remain open (6, 8, and 16). Problem 6 asks for a rigorous treatment of the axioms of physics; considering the great progress physics has made since 1900, it is clear that what would have satisfied Hilbert then would not be satisfactory now. Problem 8 is the Riemann hypothesis, and problem 16 demands a qualitative but exhaustive description of the kinds of curves that can arise from real algebraic functions. For those interested in more details, see Benjamin Yandell's (2002) marvelous book. A more technical description (a little out of date) is in Browder (1976).

The problem of graphical display that now affects us most profoundly is how to represent complex multidimensional data on two-dimensional surfaces, what Edward Tufte characterizes as escaping flatland. If we wish to examine the relationship between weight and Type II diabetes, it is easy to draw a graph. But how do we draw the graph if we also wish to include age? Drawing a three-dimensional graph is more difficult but still possible. But suppose we also wish to include amount of exercise done? And what about dietary variables? What about smoking? Race? Geographic location? Sex? Obviously, we would want to include many things for a realistic representation of the scientific issue, but our two-dimensional Euclidean perceptions get in the way. How indeed can we escape flatland?

These problems have been with us for a long time and there have been a number of enormously clever solutions (remember Galton's multidimensional weather charts discussed in chapter 8). But most of these solutions use icons to represent the data, which require us to read the graph rather than just "see" it. Reading a graph takes time and learning; seeing takes place in an instant. We need no instruction to comprehend the meaning of the soaring upward swoop of the line representing the Dow Jones Industrial Average in the 1990s (chapter 14). Can we develop such evocative displays for complex data?

Modern data graphers have available to them powerful cyberassistance that can do many things easily that could not have been accomplished at all in the past. How can this power be harnessed?

In the next three chapters I discuss some tools that will surely help. Chapter 19 is a short biography of John Wilder Tukey (1915–2000), a Princeton genius, who for many years was the person to whom very smart scientists from many fields could turn when they were stumped. Chapters 20 and 21 describe the result of a series on conversations I had with Tukey on the topic of harnessing the power of the computer to represent graphically multidimensional data in an evocative way. I have faith that the inventions Tukey laid out for future graphics will be as influential on twenty-first-century practice as Playfair's inventions were on the graphics of the nineteenth and twentieth centuries.

19 John Wilder Tukey: The Father of Twenty-first-Century Graphical Display

Our ability to see patterns in data has been improved by the light shed by a number of brilliant contributors. I discussed some of the early stars in part I, with a focus on William Playfair, whose contributions dominated all others of his time. The twentieth century, too, had important contributors to the growing science of effective data display, but John Wilder Tukey (1915–2000) stands out. Moreover, just as Playfair's work, completed during the cusp of the eighteenth and nineteenth centuries, carried over to have its greatest effect in the twentieth century, so too I suspect will Tukey's carry forward into the twenty-first and beyond.

> We have watched at least four presidents of the United States listen to him and heed his counsel.
> —William O. Baker, Retired Chairman of the Board, Bell Labs

John Tukey was a member of the President's Scientific Advisory Committee for Presidents Eisenhower, Kennedy, and Johnson. He was awarded the National Medal of Science by Richard Nixon and was special in many ways. He merged the scientific, governmental, technological, and industrial worlds more seamlessly than anyone else in the twentieth century. His scientific knowledge, creativity, experience, calculating skills, and energy were prodigious. He was famous for creating both new statistical concepts and new words to describe them.[1]

Tukey was born in New Bedford, Massachusetts, on June 16, 1915, the only child of Ralph H. Tukey and Adah Tasker Tukey. His mother was the valedictorian of the class of 1898 at Bates College and her closest competition, the salutatorian, was her eventual husband. John's father became a Latin teacher at New Bedford's high school but, because of a nepotism rule that forbade wives from working in the same school as their husbands, Mrs. Tukey worked as a private tutor, focusing especially on her son, who she recognized early on was a prodigy. Thus John was largely homeschooled.* He matriculated at Brown University, completing bachelor's and master's degrees in chemistry in 1936 and 1937, respectively.† He enrolled the following year in Princeton's graduate school. He began in chemistry but transferred during his first year to

The material in this chapter has been taken from many sources. Most prominent among these are the two biographies written by two of Tukey's long-term friends and former students, Frederick Mosteller and David Brillinger. I am grateful for their permission to use their work here.

* He attended a few high school classes regularly (such as French) that his mother did not feel qualified to teach. He also used the New Bedford library extensively.

† He chose chemistry for a mixture of reasons, but one was that he found that he could understand the chemistry journals in the New Bedford library, but not the mathematics journals. He remedied that inability during his time at Brown.

mathematics.* He completed a master's degree in 1938 and a doctorate in 1939 in mathematics.†

Princeton University, which usually does not hire its own graduate students, did not achieve its present high status by letting the dogmatic application of such policies get in the way of retaining someone of Tukey's obvious talent. Thus Princeton appointed him as an instructor in the Department of Mathematics in 1939. He stayed at Princeton for the rest of his life. During the six decades of his professional career, he accumulated many honors; among them: he was a member of the National Academy of Science, a recipient of the National Medal of Science, and the holder of an armful of honorary doctorates. He was also an avid folk dancer, bird watcher, mystery novel fan, dedicated husband, and the best mentor any graduate student could ever hope for. He was also scrupulously honest in both his scientific research and his life. Thus in many ways, he seems to lack the sorts of character traits present in William Playfair that we have argued (in chapters 1, 2, and 3) were crucial in the person who invented graphics. Yet Tukey did more than merely invent and popularize new graphical methods; by his very participation in this work, he also made the study of graphical procedures an honorable intellectual pursuit. How did this come to be? Aristotle, in his *Metaphysics*, pointed out that "we understand best those things we see grow from their very beginnings."

His Very Beginnings

As it did for most people alive during that time, World War II had a profound effect on Tukey's career. He was recruited to join Princeton's Fire Control Research Office, which did applied work on ways to make weapons more effective. Tukey thus came into close contact with a community of some of the best statisticians in the country. The recommendations they were asked to make were based on data. Within this milieu Tukey found a home. He enjoyed the work and created a torrent of inventions that would allow them to accomplish their tasks better. At the end of the war he shifted his academic focus to the statistical wing of the Department of Mathematics and, at the same time, joined AT&T's Bell Telephone Labs.‡ He was to maintain these two connections for the next fifty years.

Tukey's contributions span many areas; his collected papers alone would fill fifteen large volumes (eight of which have already been published). Some of these contributions, such as the fast Fourier transform, invaluable in spectral analysis, are too technical for inclusion here. Others, such as the stem-and-leaf diagram, are so broadly useful that they have even become part of high school mathematics curriculum. Happily, he had enough to say about how to display data that the narrow focus of these chapters will not require them to be overly parsimonious. Indeed, I can only touch on his work in this area.

One of Tukey's principal contributions was the legitimization and enrichment of what he called "exploratory data analysis." He broke data analysis into two parts, which he characterized as exploratory and confirmatory. Confirmatory analysis he described as being

* Tukey referred to this decision as "falling over the fence."

† His thesis, *Convergence and Uniformity in Topology*, was so outstanding that it was published as a book by the Annals of Mathematics Studies (Tukey 1939), only the second time this had ever been done.

‡ His joining Bell Labs was very much an extension of his Fire Control work. Tukey was hired at Bell Labs by Hendrik Bode, head of military systems and best man at his wedding, to codesign the general elements of Nike, the world's first antiaircraft (surface-to-air) missile. Military work was very much John's wartime entrée to Bell Labs, although he did many other things there in the fullness of time.

essentially judicial in nature: in it, evidence is weighed and a decision is reached. Most of the inferential statistics in the fifty years before the publication of Tukey's *Exploratory Data Analysis* (1977) concern themselves with this aspect. Exploratory analysis is more like detective work: in it, data are examined and clues are uncovered. This is where the action is. Somehow we must examine data and look for suggestive patterns. Tukey thought that it is rare that numbers do not tell you something.*

Finding unexpected patterns is a path to discovery, and Tukey provided copious evidence that drawing a picture is the best way to do it. He also invented an unending number of novel kinds of displays.

To illustrate how a simple idea of Tukey's can help lead to unexpected discoveries, let us consider a small data set and see what we can learn from it. Table 19.1 lists the homicide rates for a selection of fifteen representative states (for 1997) as well as the lowest temperature ever recorded in each state's capital.† The data table is arranged alphabetically. What can we tell from such a table? Having Alabama first is rarely on the road to insight. For a first look, the stem-and-leaf display is just the thing.

What, you ask, is a stem-and-leaf display? It is a way of jotting down a batch of numbers quickly into a form that allows the flying eye to see patterns. To construct such a display we first note that homicide rates range from a low 1.8 per hundred thousand in Maine, to a high of 16.1 in Louisiana. So we

Table 19.1
Homicide Rates per 100,000 Population for Fifteen States (1997) and the Lowest Temperature Ever Recorded in Each State's Capital (Degrees F)

State	Lowest temperature	Homicide rate
Alabama	7	12.0
California	20	8.8
Connecticut	-26	3.9
Georgia	-3	8.7
Illinois	-25	9.8
Kansas	-12	6.1
Louisiana	14	16.1
Maine	-39	1.8
Maryland	-7	10.9
Minnesota	-34	2.8
Mississippi	6	14.2
Nebraska	-22	3.9
New Hampshire	-37	2.2
New York	-28	6.3
South Dakota	-36	3.0

Source: Data from ***Statistical Abstracts of the United States, 2000*** (Washington, DC: Government Printing Office, 2001).

write down a "stem" of sixteen numbers (follow along on p. 120) to represent the integer part of these data.

Next we run down the list of states from table 19.1 and add the tenths entry for each state to the right of the appropriate stem. These are the "leaves." So for Alabama we add a 0 next to the 12, for Connecticut a 9 next to the 3, and for Illinois we add an 8 next to the 9. Because we are writing down only one digit, this can be done very quickly, and soon an evocative picture emerges. The middle diagram on page 120 is a stem-and-leaf plot for the

One Effect of the AIDS Epidemic

	Share price			Percentage gain since 1985
	1985	1990	1997	
Club Med	$21.00	$21.00	$14.50	−31
Pacific Dunlap*	$1.60	$4.50	$10.80	575
S&P 500	185	340	937	406

*Pacific Dunlap is a major condom manufacturer.

* Consider, for example, the suggestive table to the right, showing what happened to the share prices of two companies between 1985 and 1997.

† Why should we adjoin data on temperature with that of homicides? Bear with me.

Stem	
16	
15	
14	
13	
12	
11	
10	
9	
8	
7	
6	
5	
4	
3	
2	
1	

Stem	Leaf
16	1
15	
14	2
13	
12	0
11	8
10	459
9	025558
8	1778
7	4667
6	1356
5	3
4	0112444577
3	01699
2	23588
1	8

States	Stem	Leaf
LA	16	1
	15	
MS	14	2
	13	
AL	12	0
AR	11	8
MD, NV, TN	10	459
AK, NC. SC, NM, AZ, IL	9	025558
MO, MI, OK, GA, CA	8	1778
IN, VA, TX, FL	7	4667
KS, NY, PA, KY	6	1356
WV	5	3
HI, WI, OR, MT, WY, OH, DE, NJ, WA, CO	4	0112444577
SD, UT, ID, NE, CT	3	01699
NH, MA. IA, RI, MN	2	23588
ME	1	8

forty-seven states on which data are available. We immediately see that states seem to fall into three groups. The first group has homicide rates between 1 and 5 per hundred thousand, the second group between 6 and 10, and the third group, consisting of four states, has the most extreme homicide rates (all greater than 11 per hundred thousand).*

Simply by writing down the numbers in an organized fashion we have seen structure that we might not have expected. But we can go still further. By identifying each state, we are likely to gain some insight into the character of homicide. We shall at least gain insight into the character of some of the states. This can be done simply within the same graphical format using a back-to-back stem-and-leaf diagram and writing a state identifier in its location (above).

Now we discover that four southern states—Louisiana, Mississippi, Alabama and Arkansas—have the highest homicide rates, and the states with the lowest homicide rates (Maine, Minnesota, Rhode Island, Iowa, Massachusetts, New Hampshire) are all northern. Is this just a result of the extremes? Or is it generally true that the colder the weather the lower the homicide rate? If we plot the data in table 19.1 as a scatterplot (figure 19.1) we see that this is a general pattern, broken only by Hawaii, whose balmy climes would suggest great danger yet somehow it is different.†

Country	Homicide rate
Iceland	0.4
Ireland	0.6
Japan	0.6
Norway	0.8
Austria	1.0
Germany	1.1
China	1.4
Israel	1.4
Canada	1.7
Czech Republic	2.1
Poland	2.8
Croatia	3.3
Nicaragua	5.5
Cuba	7.8
Belarus	10.4
Mexico	17.2
Brazil	19.0

* To place these numbers in perspective, note that the homicide rate in Canada during this same year was 1.7, and Israel, which was suffering through the beginning of the intifada, was only 1.4. Viewing a sampling of homicide rates from around the world (from the *United Nations 1996 Demographic Yearbook*, published in 1998) provides a context.

† This is hardly the first time that the relationship between warm weather and mayhem has been noticed. An early and particularly enlightened discussion of this topic is in Charles Angoff and H. L. Mencken's delightful essay "The Worst American State" (Angoff and Mencken 1931). It is well worth reading, not the least for their causal analysis of this effect, which, as I recall, has something to do with the excessive number of Baptists in the South.

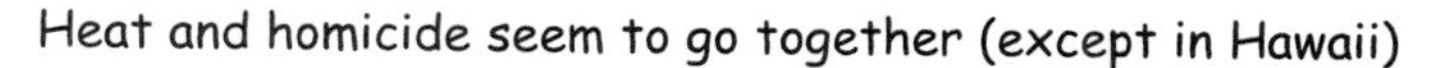

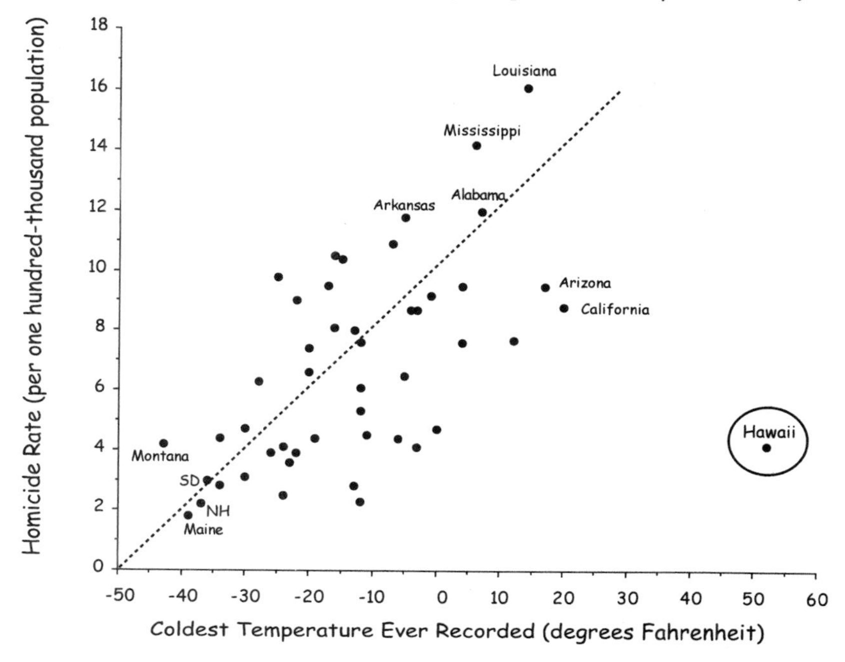

Figure 19.1. A plot comparing the homicide rates in each state with the coldest temperature ever recorded in that state shows a dramatic relationship between warmth and homicide (except for Hawaii).

This brief excursion into exploratory data analysis provides a glimpse into the character of one of Tukey's contributions. He made it acceptable for scientists to ply their craft by constructing simple pictures and, while keeping an open mind, searching for suggestive patterns. If such a pattern seems to emerge, the next step is always to gather ancillary information and, almost surely with new plots, look for confirmation of the interpretation of the original pattern. This notion merges with a great Aristotelian theme—the search for relations between things apparently disconnected. It has often been found that when we find "similitude in things to common view unlike" we can turn old facts into new knowledge. "Newton did not show the cause of the apple falling, but he showed a similitude between the apple and the stars."[2] Newton was content to be able to bring many diverse phenomena under a few principles, even though the cause of these principles remained undiscovered. Both David Hume and John Stuart Mill declared that all reasoning whatsoever depends on resemblance or analogy, and the power to recognize it.

Tukey's other contributions, similarly, share some important characteristics: they are only as complex as required by the task they are designed to perform; they avoid pomposity.* And they are broadly useful and

* In addition to being a fine mathematician, Tukey was also a word man. He felt that if you invent something, you need to also invent a word for it so that your idea does not get confused with something similar but in some important way different. Tukey's names were always simple and evocative. If he had been in charge of naming plants or animals, the current typology of Latin would never have occurred. *Stem-and-leaf* is one example; some other, better-known terms of his are the computer terms *bit* (for binary integer) and *software* (as distinguished from the hardware of the computing machine). Lesser-known words are *jackknife* (for a statistical method that works for just about anything), *ANOVA*, shorthand for Fisher's Analysis of Variance, *hinge*, *trimeans*, *trimming*, and dozens of others.

broadly trustworthy since they do not rely, for their validity, on invisible assumptions. The analysis I did here also conveys, in a small way, the continuing process that Tukey envisioned most data analysis following. A preliminary look (the stem-and-leaf of homicide rates) suggests looking for ancillary information to help explain the observed structure. Adding state names to the homicide rates gives further hints, and the plot of weather information confirms these initial indications of the systematic geographic nature of the systematic variation of homicide. It also identifies Hawaii as unusual. No analysis is ever complete, but each step moves us along the pathway toward fuller understanding.

John Tukey and William Playfair shared many characteristics. They were both empiricists down to the bottom of their souls. They both had quicksilver intelligence,* and a positive attraction to solving practical problems. Like Playfair, Tukey was an iconoclast, although he manifested this in more socially acceptable ways than blackmail and extortion. This willingness, indeed desire, to go against the intellectual mainstream if that seemed sensible (or perhaps if it was not too nonsensical), manifests itself throughout his corpus of work. In graphics it was the very act of spending effort in an area that others felt was unworthy of consideration of a serious scholar (remember Luke Howard's apology in chapter 1). But even in more mathematically technical arenas his puckishness would often manifest itself.†

Playfair and Tukey: Three Points of Agreement

In the more than two hundred years since the initial development of graphic techniques, we have acquired some wisdom in their use, borne of experience. It is a telling measure of Playfair's accomplishment to note how many points of agreement there seem to be between Tukey and Playfair on important aspects of graphical display. Among these are:

1. Impact is important

> Along with Playfair's desire to tell the story of history graphically was the desire to tell it dramatically.
> (Costigan-Eaves and Macdonald-Ross 1990)

> The greatest possibilities of visual display lie in vividness and inescapability of the intended message.
> (Tukey 1989)

* At a commemorative celebration of Tukey's seventieth birthday, and hence his formal retirement from Bell Labs, the speakers often referred to how his special genius aided them in their own work. This was especially impressive considering the credentials of those speakers (including at least one Nobel laureate). Mark Kac, the noted mathematician, refers to two kinds of geniuses: the ordinary genius, whose work we could have done if we were just many times smarter, and the magician, whose mind "is in the orthogonal complement of ours" and even after we learn what he has done, we have no idea how he did it. Tukey was a magician of the highest order.

† In 1959, Tukey journeyed to Geneva, Switzerland, where he served as a U.S. delegate to the Conference on the Discontinuance of Nuclear Weapon Tests. The technical problems associated with distinguishing between a nuclear explosion and an earthquake are of obvious importance for enforcing any ban. Tukey observed that when a signal was laid on top of itself after a delay (which might happen when the energy from an underground explosion radiates downward until striking some solid surface and rebounds upward), there is a ripple in its spectrum that can allow us to estimate the depth of the source of the signal. Because subterranean nuclear tests are typically shallower than most earthquakes, this provided an important tool for detection of treaty violations. In his 1963 paper on the underlying technology for analyzing the frequency data (Bogert, Healy, and Tukey 1963), he emphasized the importance of signal reflections by choosing a reflected title, "The Quefrency Alanysis of Time Series for Echoes," and the task of examining phase shifts in the spectrum he termed "saphe cracking." I assume that I was not the only reader who corrected typos for two pages before realizing that he really meant it that way.

2. Understanding graphs is not always automatic

> Those who do not, at first sight, understand the manner of inspecting the Charts, will read, with attention the few lines of directions facing the first Chart, after which they will find all the difficulty entirely vanish, and as much information may be obtained in five minutes as would require whole days to imprint on the memory, in a lasting manner, by a table of figures.
> (Playfair 1801, p. xii)

> A picture may be worth a thousand words, but it may take a hundred words to do it.
> (Tukey 1986)

There seems to be a widespread belief that a good graph should be entirely comprehensible without any instruction. Such a view is limiting, to be sure. It seems to me that we can divide good graphs into at least two categories:

> A strongly good graph tells us everything we need to know just by looking at it.
> A weakly good graph tells us everything we need to know just by looking at it, once we know how to look.

A good legend can transform a weakly good graph into a strongly good one. We ought to make this transformation when possible. A legend should do more than merely label the components of the plot. Instead it should tell us what is important, what the point of the graph is. This serves two purposes. First, it informs the viewer, transforming what might be a weakly good graph into a strongly good one. But second, and of at least equal importance, it forces the grapher to think about why this graph is being prepared. Insisting on informative legends can substantially reduce the number of pointless graphs we see, and it will better structure the meaningful ones because once a grapher is clear about what the point of the graph is, the appropriate way to structure it becomes clearer as well.

3. A graph can tell us things easily that might not have been seen otherwise

> I found the first rough draft gave me a better comprehension of the subject, than all that I had learnt from occasional reading, for half my lifetime.
> (Playfair 1786)

> The greatest value of a graph is when it forces us to see what we never expected.
> (Tukey 1977)

We have all had the experience of seeing something in a data set once we had graphed it that had lain hidden throughout many previous analyses. This is one reason that many of us have learned to start data analysis with graphics. There are too many examples of this to try to choose a "best" one, but one of my favorites was presented by Bill Cleveland in his remarkable book *The Elements of Graphing Data*.[3]

Sir Cyril Burt published data in a 1961 paper showing the mean IQ of over forty thousand father-child pairs divided into six social classes (see table 19.2).[4]

These data look innocent enough when tabled, but when graphed they are far too linear for us to trust. In fact, they match, virtually perfectly, a simple equation that Jane

Table 19.2
Mean IQs of Fathers and Children in Six Social Classes

Occupation Category	Mean IQ	
	Father	Child
Higher Professional	139.7	120.8
Lower Professional	130.6	114.7
Clerical	115.9	107.8
Skilled	108.2	104.6
Semiskilled	97.8	98.9
Unskilled	84.9	92.6

Source: Burt 1961.

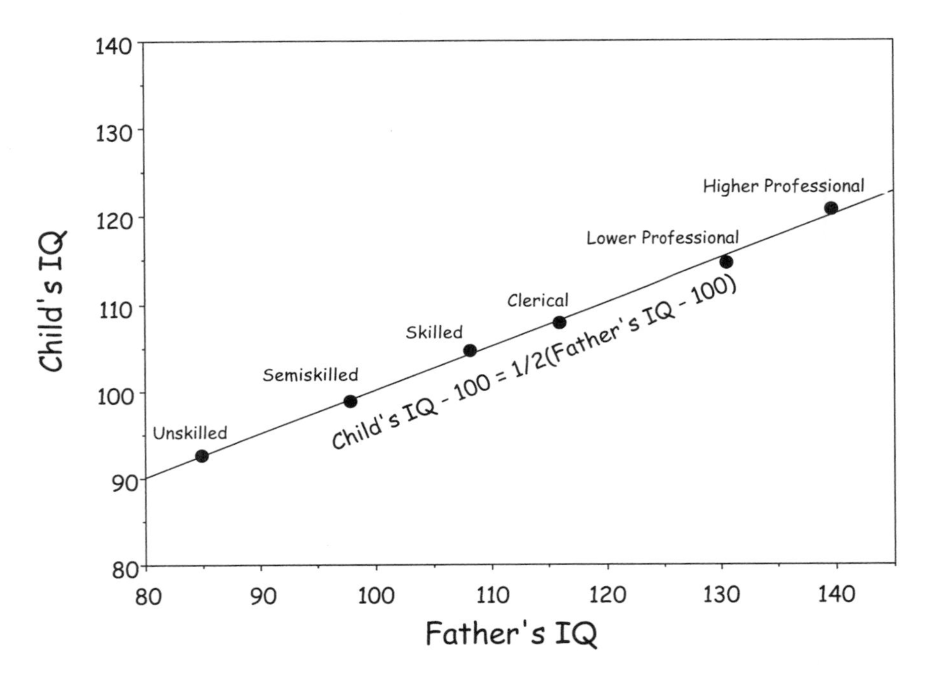

Figure 19.2. Cyril Burt's table of the average IQs of fathers and children divided by father's occupation seems reasonable until it is graphed. Then we can see that it may be too linear to be true (but see Stigler 1979 for an alternative explanation).

Conway had offered as a way of predicting a child's IQ in a particular social class from a father's IQ in that class:[5]

(Child's IQ – 100) = 1/2 (Father's IQ – 100).

It wasn't until 1978, in an article by Donald Dorfman, that this was challenged.[6] Yet had anyone graphed it, its validity would have been questioned immediately (see figure 19.2). A careful biography, published after Burt's death,[7] showed that not only had Burt fabricated the data, he also wrote the article nominally authored by what turned out to be an imaginary Conway. He also edited the journal in which all of this appeared.

The graphical tools that Playfair prepared for us, which were augmented by much of Tukey's work, were designed for the dominant presentation medium of the day—paper and ink. It has been remarkable just how well we have been served despite the two-dimensional limits of such a medium. Yet real problems often have an irreducible complexity that defies two-dimensional representation. There have often been imaginative solutions to "escape flatland" (to use Tufte's evocative phrase). But the twenty-first century brings with it powerful, cheap computers that provide great potential for different kinds of displays, using tools that have not yet been devised. In the rest of this part I will describe a broad class of tools that Tukey believed would provide a beginning. These were, quite literally, the very last things that Tukey did.

20 Graphical Tools for the Twenty-first Century: I. Spinning and Slicing

It has been more than thirty years since I first stumbled into the University of Chicago's Green Hall and began teaching a course in exploratory data analysis (EDA). A great deal has changed since then, but a great deal remains the same. Although the tools of EDA at that time were primarily built for hand manipulation, we found that a computer program we wrote (even a batch processed mainframe program) was enormously helpful.[1] Computing has changed profoundly in those intervening thirty years, but most of the EDA tools that John Tukey laid out in his wonderful 1977 book remain broadly helpful. Indeed, automating those tools made them more accessible and more helpful than ever before. But with the added power of modern computing, what new EDA tools might be useful?

Extrapolation is always a pretty dicey proposition, but it is pretty certain that variations on existing tools are certainly going to be among those that will find their way into our future. But what new tools, now possible with cheap and powerful high-speed computing, may also be useful? In the year before his death, I had a series of conversations with John Tukey that focused precisely on this topic. Because Tukey's eye toward the future has always been uncanny, I have considerable confidence in his projections.

The future of EDA will resemble the past in that useful new tools will have to be chosen with wisdom and discipline. The wisdom will be born of experience in data analysis and, with it, an understanding of the underlying variability that is the hallmark of all stochastic processes.* And the discipline is necessary to avoid procedures that do not work very well despite their flashy nature. The commonly seen pseudo-three-dimensional color pie chart comes to mind as an obvious example of the triumph of flash over experience. Discipline to avoid such excesses is going to be more important in the future as the tools of computer presentation and dynamic display present us with even more opportunities to communicate confusion more effectively.

Communicating confusion brings me to *data mining*, a currently trendy term referring to a wide mix of largely graphical procedures, many of which seem to have multicolored

* The measurement of anything will yield variability. In some deep sense we all know this, but it is easy to forget. We know that the average SAT scores from this year's senior class at the local high school will not be the same as they were last year, and we do not consider firing the principal if they drop a little. We understand that even if everything remained the same, we should expect some variation. Yet if we have a powerful data analyzer in our hands, it is easy to isolate artifactually unusual points and assign importance to them. Even Nobel Prize winners (remember the story of W. F. Sharpe, told in chapter 10) can fall prey to it.

towers arrayed in perspective on a grid of trapezoids. To fully understand data mining, one has to retreat more than a century to the psychophysics labs of the German Gestalt psychologists. They developed an experimental procedure they called the "Ganzwelt," in which the subject was confronted with an absolutely uniform visual field—roughly what you would see if you had half a Ping-Pong ball glued over each eye. They found that after you stared at absolutely nothing long enough, you would start to see patterns. So it is with data mining.

The future of real EDA will undoubtedly involve the computer in highly interactive analyses, and so we must look for what sorts of tools are likely to prove helpful in such an environment. In this chapter I report on two analysis engines that I believe are helpful additions to every analyst's toolbox. Because these engines are specifically designed to operate within a highly interactive environment, they are different from traditional EDA tools,[2] which will surely remain valuable.

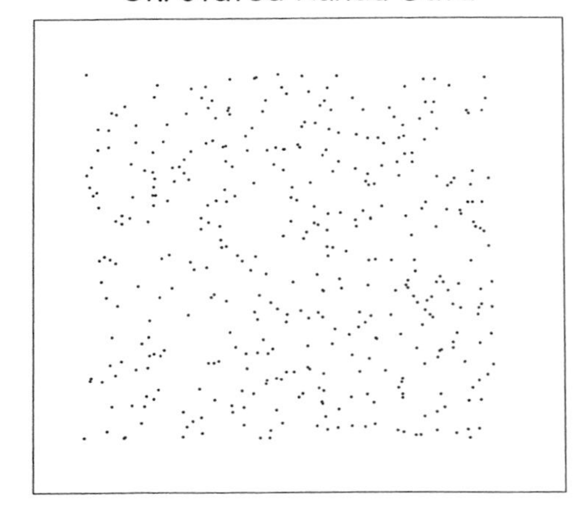

Figure 20.1. A two-dimensional projection of four hundred points plotted in the unit cube generated by the congruential random number generator Randu.

Highly interactive computerized data analysis allows us to use motion to convey information. Although motion is possible without a computer (flipping pages with slightly varying versions on each page), it has been too cumbersome in the past to be anything but a novelty. But now motion is easy and we need to understand how to use it.

To begin, let me make a distinction that will become important in subsequent discussions. I shall call a display "mobile" if the goal is to manipulate the data and the display in such a way as to arrive at some revealing, but static, view. I shall call a display "dynamic" if the revealing characteristics are only available while the display is in motion—the message is in the voyage. Because of the obvious practical restrictions, this chapter will focus on mobile displays more than dynamic ones. This does not reflect my estimate of their relative future worth.

An interesting illustration of the value of a mobile display comes from an investigation of IBM's ill-fated random number generator, Randu. Modern science uses random numbers in many ways, most often for powering simulations of various phenomena whose stochastic character requires a random component. Thus it is important to have a way to generate numbers that are as close to being truly random as possible. Randu was IBM's offering. It is of a linear congruential type that turns out to yield numbers that depart from randomness in an interesting way. Suppose we produce 1,200 numbers between 0 and 1 with this generator and consider each succeeding triple a point in three-dimensional space. We should end up with a uniform distribution of points on the unit cube. A two-dimensional projection of that cube is shown in figure 20.1. Nothing in this display looks out of the ordinary. It looks like a scattering of four hundred random points all over the two-dimensional projection. This is just as we would expect.

But as we sit watching the three-dimensional cube of random points spin before our eyes, something remarkable occurs. The image of the slowly spinning random points suddenly shifts and all the points line up on fifteen planes in three-dimensional space!* See figure 20.2. This is a most decidedly nonrandom configuration. We note that this pattern of fifteen stripes disappears as quickly as it came. Indeed, if we rotate away from this viewpoint by even a few degrees, the image of randomness returns. This phenomenon is familiar to anyone who has ever driven past a cornfield and noticed how the cornrows sometimes line up and at all other times look as if they are planted helter-skelter.†

George Marsaglia first described this flaw in Randu in 1968 in his lyrically titled paper, "Random Numbers Fall Mainly in the Planes," but it is trivially uncovered with a rotation engine. The story would be more dramatically told if it were done dynamically, but the value of the outcome can be fully appreciated with the static view of the end result.

The character of a written document effectively precludes the inclusion of a live dynamic example, so instead let me ask you to use your imagination and place yourself back

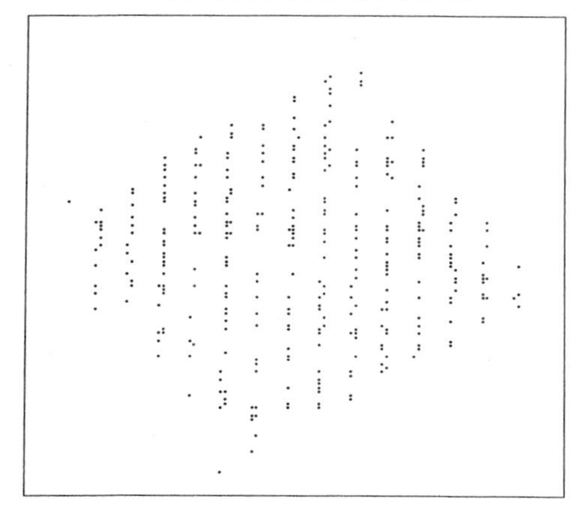

Figure 20.2. A different two-dimensional projection of the cube shown in figure 20.1 showing the striped pattern that is evidence for the conclusion that Randu does not yield entirely random numbers.

in time about four hundred years, to 1605 in Padua. You are a young assistant to the eminent professor of mathematics Galileo Galilei.[3] He is feeling frustrated, for his astronomical observations have reached a dead end because of limitations in his observational equipment and he is faced with a difficult decision. Should he continue the construction

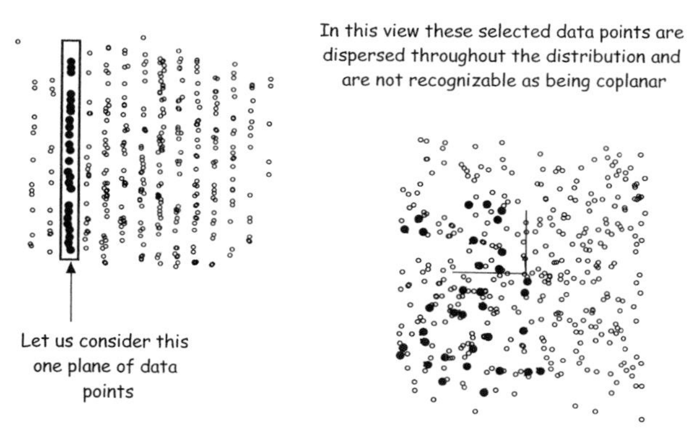

Figure 20.3 (left). The striped pattern from figure 20.2, in which the thirty-four points on one plane are highlighted.

Figure 20.4 (right). After highlighting thirty-four points on one plane, we rotate the entire data cube a few degrees, so that it is now close to the original configuration. The highlighted coplanar points no longer show any relationship to one another.

* It is now well known that all congruential random number generators have this flaw; they all generate points that line up on p-planes in k-space. It would not have been as serious if Randu had been built so that it yielded points on 1,634 planes in a 98-dimensional space, but when p is 15 and k is 3, the resulting numbers are too granular for many purposes.

† This point can be made dramatically by spinning backward. Consider the aligned version of the Randu points shown in figure 20.3, in which I have highlighted the points in one of the planes by marking them with filled-in dots. When I rotate the configuration even just a few degrees (figure 20.4) we lose all sense that those points have any relation to one another.

Table 20.1

Height (in steps)	Water (in thimbles)
5	0.6
10	0.8
15	1.0
20	1.1
25	1.3
30	1.4
40	1.6
80	2.2
100	2.5
200	3.5

of a calculating machine to aid in analyzing the astronomical data? Or should he focus his attention on improving the performance of his refracting telescope? He figures, correctly, that without data his analysis machine is worthless, so he has turned his attention to the telescope and handed his data and the almost-finished calculator over to you with the instructions, "see what you can make of our observations."

The "observations" he is referring to come from an ingenious experiment you helped him run over the past several weeks, in which a heavy object was slid down a long inclined plane from a fixed number of steps up a tower. When it was dropped, water was allowed to run out of a tube into a container. When the object landed, the water flow was quickly stopped and the amount of water in the container was measured. This was repeated at ten different heights. Your data are shown in table 20.1.

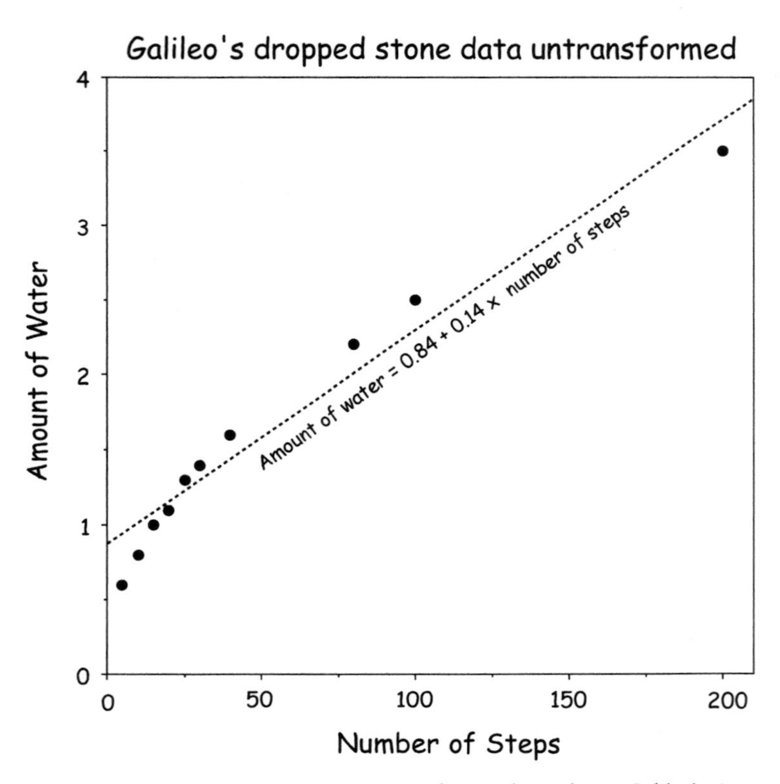

Figure 20.5. A plot showing the amount of water drained into Galileo's cistern as a function of the height of a dropped stone. A best-fitting straight line is drawn in, showing the curvature of the data points.

Using Galileo's computing device, which is being cranked by a visiting nun who claims some familial relation to the professor, you input these data and use the analysis program that Galileo calls "Copernicus." The program immediately plots the points (see figure 20.5), drawing a straight line through the plot to act as a guide. You can see that the higher the drop-point was, the more water flowed, but that is hardly a scientific breakthrough. You also note that the points from greater heights were the ones that were the principal cause of the lack of linearity. "Copernicus" suggested, "Try transforming the amount of water." Transform, sure, but how? "Copernicus" suggested, "Raising it to some power will surely affect the curvature of the points," and a slider appears on the screen beneath the plot of the data points. The arrow is pointed at 1.

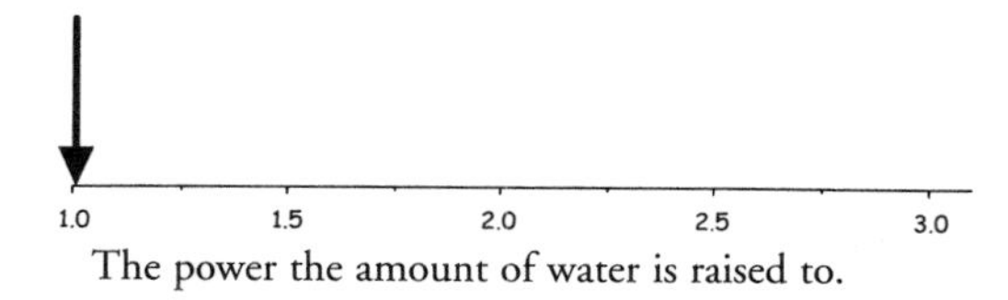

You grab the arrow and move it to the right. As you do, the vertical axis changes its size and the points in the plot straighten and then, when the arrow is at 3, their curvature is clearly reversed (see figure 20.6).

"Aha," you think, "if it is curved downward when I raise it to the first power (untransformed) and upward when it is raised to the third, it must be straight somewhere in between." Your hopes of finding a simple answer are bolstered and you begin to move the slider slowly and see that at around 2 the points are almost perfectly straight. Happily, you notice that the difference in the amount

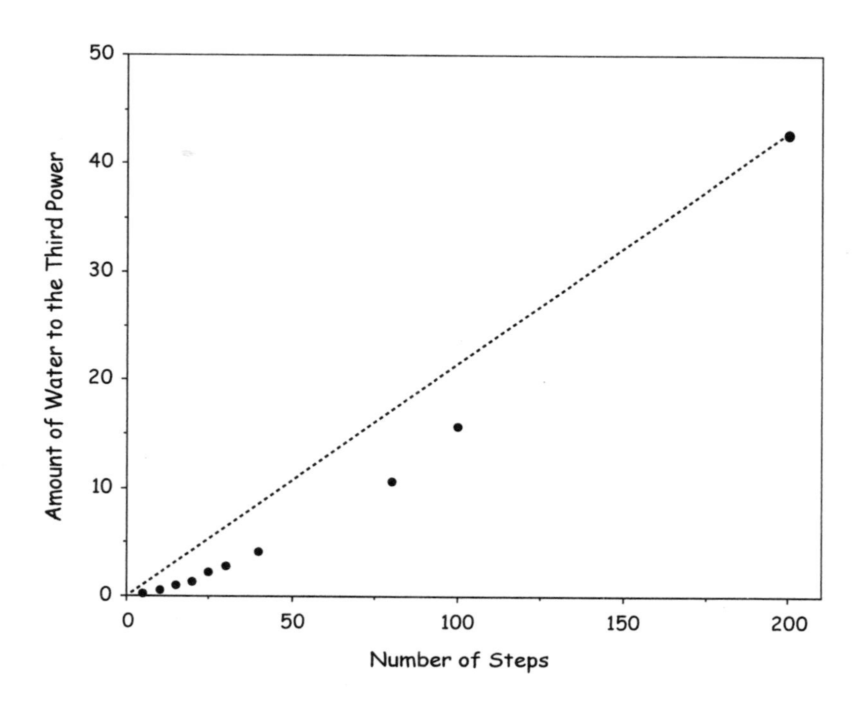

Figure 20.6. The same data as in figure 20.5, but the values associated with the amount of water were raised to the third power. The curvature seen in figure 20.5 is reversed.

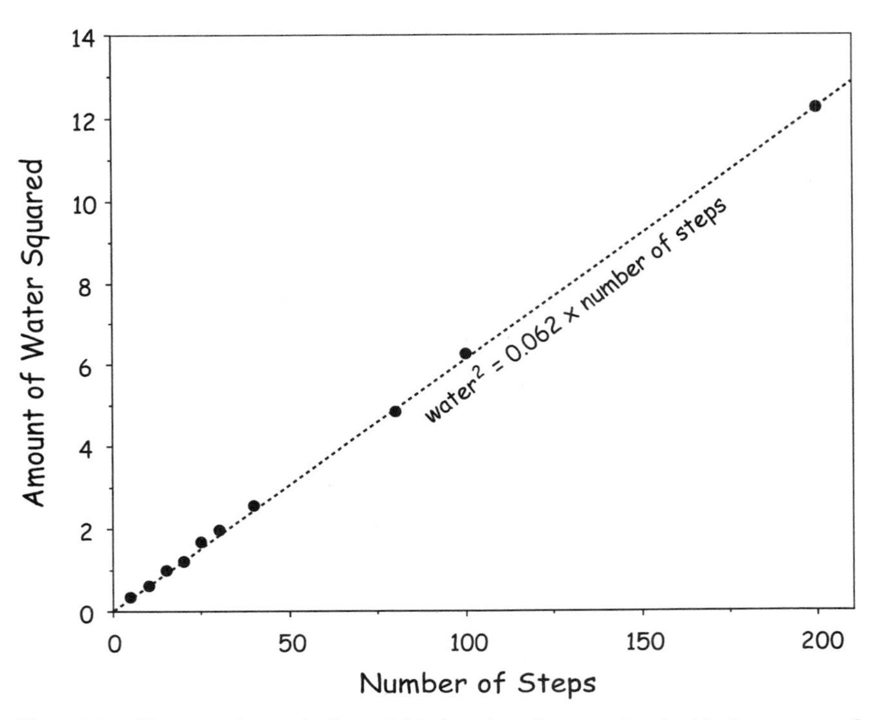

Figure 20.7. The same data as in figure 20.5, but the values associated with the amount of water were raised to the second power. The curvature seen in figure 20.5 is gone. The fitted linear function can be easily manipulated to yield one of Newton's Laws.

of straightness between 1.8 and 2.2 is slight and so even if 2 is not the right answer it will not do you much harm. At this point, Sister Maria Celeste, the nun providing power for your calculations, politely asks if your analyses are almost complete, as her arm has gotten quite numb and she is about to pass out from exhaustion. So you note down the equation superimposed over the plot when the slider is at 2, close the analysis window, thank Sister Maria Celeste, and retire to your desk for further thought.

The final result is shown in figure 20.7, and the equation can be easily generalized. We know that the amount of water is directly related to time (say, t), and the number of steps is just a proxy for the distance fallen (say, d), and so we can phrase the final result as

$$t^2 = .062d.$$

Or, if we wish to know how far something will fall if we drop it for t, we arrive at the familiar equation*

$$d = 16t^2.$$

What has this bit of historical fantasy taught us? We were investigating the effect of dynamically reexpressing a variable (*amount of water*) in an equation. We built a linear model to include a dynamically alterable vari-

* Familiar at least if we calculate time in seconds and distance in feet. But it will have to await the next century and Newton's *Principia* to be fully explicated. It is daunting to see how lucky Newton was, for if Galileo had a little more EDA (and a little less ecclesiastical censure), Newton's Laws of Motion might now be called Galileo's Laws. The laws were there in Galileo's data, just waiting for him to discover them.

able that reexpresses one variable to the pth power. We then slid the value of p dynamically between 1 and 3 while watching the displays change smoothly.

We saw:

1. How the reexpression affects the fit to the straight line as well as what value of p seems best,
2. Whether any points become outliers or lose their status as outliers because of the reexpression,
3. How sensitive the model is to choice of reexpression, especially in the vicinity of the best choice of p.

In particular, (3) provides information about the path in the vicinity of good solutions and lets us see immediately whether forcing p to a "nice" value like 2 has bad consequences, and whether the decision to reexpress at all has far-reaching consequences.

I hope that this provides some, albeit static, evidence in support of the worth of dynamic displays. The value of dynamism is not merely the final destination, but also the path taken to it.

Let us next consider some ways that we can use dynamism in plotting.

Tool 1. Rotatable Scatterplots

Plotting points provides us with insights when the points displayed look like something familiar. What do we mean by "familiar"? If they form clumps and the points within each clump seem to share some important characteristic, we have an insight. If they fall on a line, that tells us something about the relationship between the variables that are carried in the points. If they follow the shape of some smooth curve or surface, we again have learned something. Three-dimensional points plotted as a sequence of projections on a two-dimensional surface only rarely reveal any three-dimensional structure they may have unless the sequence of two-dimensional projections is chosen carefully and they are shown at an appropriate speed. One way to do this is to make the points appear to be rotating through their third dimension while being displayed in the other two.* It was exactly this sort of procedure that made our discovery of Randu's three-dimensional flaws so easy. Of course the use of this procedure can be expanded easily into other arenas, but describing those expansions without the benefit of an accompanying dynamic display is too difficult.†

* This methodology was first explored in the computer system PRIM-9 (Friedman, Tukey, and Tukey 1980) and has since found its way into many statistics and graphics packages (MacSpin—Donoho, Donoho, and Gasko 1986; DataDesk—Velleman 1997).

† Too difficult without assuming that the reader has a substantial amount of statistical expertise. If you have such expertise, read on. One such area of expansion is to construct rotatable scatterplot matrices (SPLOMs) in which all of the scatterplots show the new projections from the 3-D point cloud that has been rotated. The rotation engine can be augmented with a "bink" button giving the sequences of transformations that yielded a particularly interesting view (for a mobile display). However, further effort is required to work out a sensible way to keep track of all the transformations that transpired so that interesting paths to discovery are not lost (for dynamic discoveries). The use of rotation within SPLOMs should be thought of as a mobile and not dynamic display. The fundamental value of this tool is for the task of exploring a point cloud of high dimensionality. The basic idea is to fly around this cloud and from each vantage point to look at the SPLOM of all pairs of dimensions from that point of view.

The goal of the exploration is to find things of interest. Some interesting things may be views that expose a point or small group of points as outliers. Finding such a result warns us about the use of summary statistics (e.g., least squares measures) that are not robust with respect to a small number of outliers. It suggests that we ought to use measures with high breakdown points.

A second interesting configuration is one that shows multiple groupings. In general, we are looking for distributions of points that seem very far away from Gaussianality.

Of course there are many paths to take when the space is highly multidimensional and so we may need some sort of analytic help in choosing paths that may yield fruit. The work associated with "The Grand Tour" (Asimov 1982) may be a place to start. But we view analytic aids to exploration as only a place to start. The next steps must be guided by what we see.

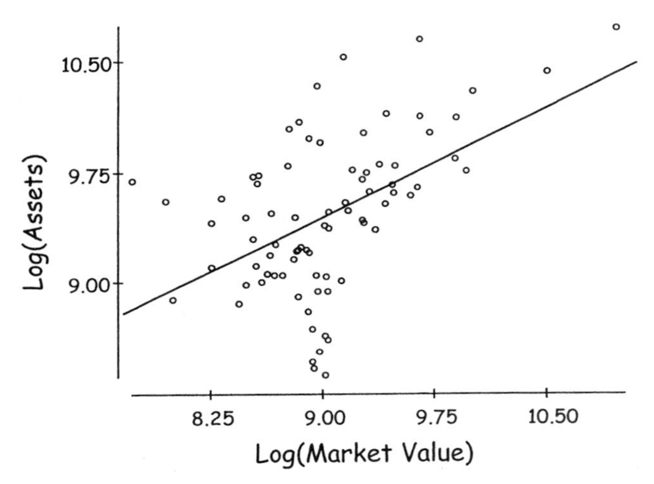

Figure 20.8. A scatterplot of Log(Assets) against Log(Market Value) for eighty companies drawn at random from the Forbes 500. A fitted regression line is drawn in.

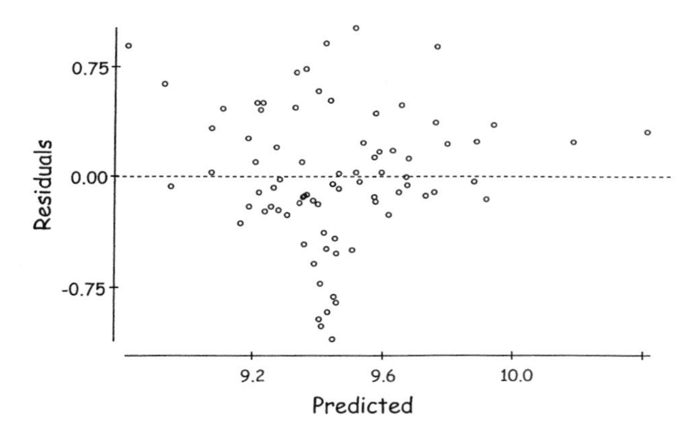

Figure 20.9. The residuals from the regression depicted in figure 20.8 are plotted against their predicted values.

Tool 2. Slicing Engine

Slicing, quite simply, is the ability to specify some screen dimension that you can move and all data points that are touched by the dimension as it is moved are highlighted/selected for special treatment. This is especially useful with linked displays in which one slices along a particular dimension in one view and watches to see where the points thus chosen show up in a very different view.

As an example, consider the scatterplot of Log(Assets) versus Log(Market Value) from eighty companies drawn randomly from the Forbes 500 (figure 20.8). We see three interesting features:

1. A trend of companies with greater market value to have greater assets (see the fitted line in figure 20.8),
2. About seven companies with a market value of about a billion dollars that have markedly lower than expected assets, and
3. A string of companies, of varying market values, that have unusually large assets.

It seems sensible to look at the vertical distance of each point from the drawn-in line (the "residuals" from the overall trend). These are shown, as a function of predicted value, in figure 20.9. We next use our slicing tool to select companies with large positive residuals (figure 20.10, panel A). When they are selected they become automatically shaded. At the same time as we select these companies, a linked bar chart, which shows the number of companies within the sample that are drawn from each of nine industrial sectors, reacts. The reaction is in real time, but a snapshot of it is shown in panel B of figure 20.10. It shows us that most of the companies with large assets relative to their market value are in the finance sector (banks).

The linking of the scatterplot with the bar chart provided the environment within which the explanatory power of slicing can be effectively used. Slicing from the bottom up shows us that companies of less than expected assets seem to be distributed more or less uniformly across all of the industrial sectors.

The slicing has also helped us in interpreting the data. Ordinarily when we find a company with greater assets than would be predicted from its market value, we suspect that we may have found a bargain. But once we see that in this instance such companies are banks, whose principal assets are tangible (rather than intangibles such as research expertise, patents, etc.), we are not surprised by their location in this graph and can thence resist the impulse to run out and buy their stock.

Slicing may not, in this instance, have told us anything that we could not have uncovered using more plebeian means, but it allowed us to uncover it *easily*. When exploring complex data sets, the ease with which a question can be answered often determines whether it will be asked at all.

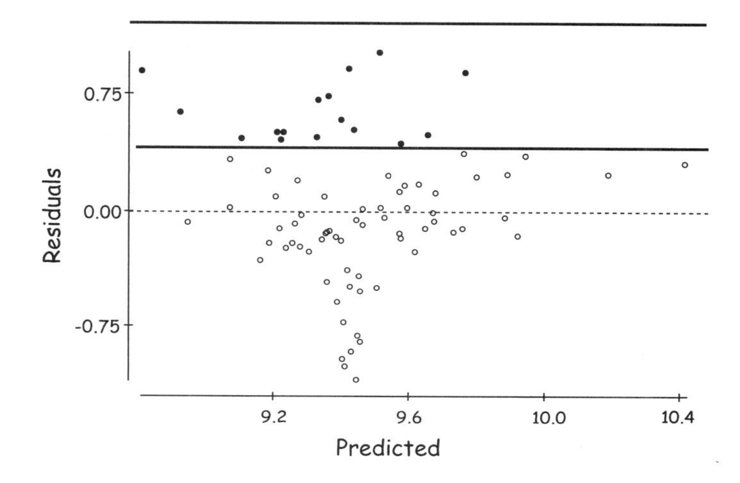

Figure 20.10, panel A. The residual plot from figure 20.9 is sliced downward from the top of the vertical axis. Those items selected by the slicer are shown darkened.

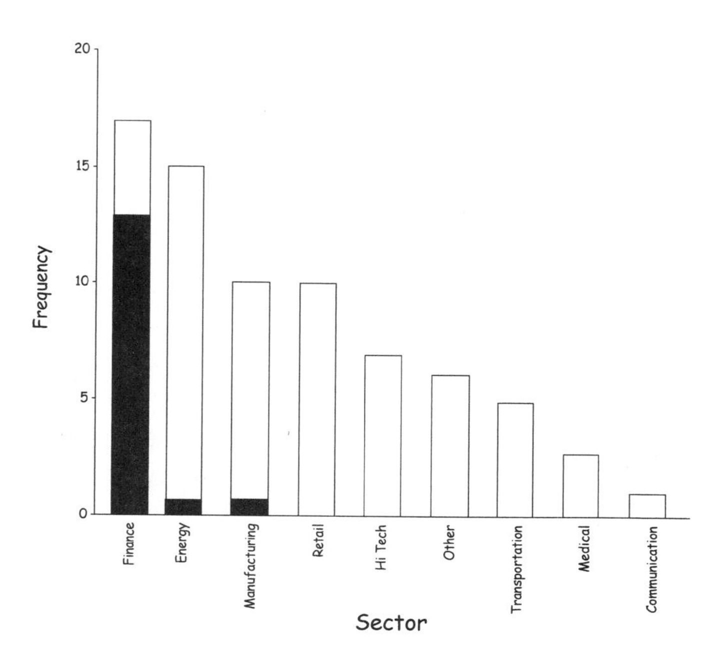

Figure 20.10, panel B. A bar chart showing the number of companies in each of nine industrial sectors. As a company is selected by the slicer in panel A, the sector that it belongs to is shaded. This display shows that most of the companies with positive residuals are in the finance sector.

21 Graphical Tools for the Twenty-first Century: II. Nearness and Smoothing Engines

The previous chapter discussed two graphical engines that can profitably be used to aid in computer-assisted graphical data analysis. I complete this discussion with two more. The four engines do not overlap appreciably in their capabilities—what one does, the others do not—but they do span a space of considerable dimensionality in their capabilities. I considered including other tools (e.g., a flashing capability, in which a sequence of points could be alternated on the screen and through their apparent movement yield information) but decided that these four, when properly implemented and coupled with automated versions of the traditional graphical tools, provide a versatile and reasonably complete toolkit that is suitable for a broad range of prospective data analysis questions.

Tool 3. Nearness Engine

A principal goal of all exploratory data analysis is to discover empirical groupings of data points. Such patterns help us to see structure that we might not have expected. Illustrations of how such discoveries have been important are easy to find; epidemiologists routinely look for geographic patterns of disease in their search for cause. One famous example is contained in a map drawn by John Snow in 1854 of deaths during a cholera epidemic in London (figure 21.1). He noticed that the deaths, marked with dots, clustered around the water pump on Broad Street (marked with an X). Using his map as evidence, he convinced the vestry of St. James to remove the handle from the pump. In less than two weeks, the epidemic, which had already taken more than five hundred lives, ended. This is often cited as the first triumph of epidemiology. The Broad Street pump is now inoperable; serving in its place is the John Snow Pub. Putting a pub in this spot is as appropriate as its name, for one of the clues that pointed Snow toward the examination of water pumps as a possible source of contagion was the dearth of cholera deaths in a neighboring brewery. We cannot know whether the principal cause of their immunity was the separate well within the brewery's walls or the company's policy of allowing employees to sample the product freely. But the anomalous safety of the brewery workers must have been one more clue for Snow.[1]

A nearness engine could help us find such groupings more easily and could thus provide the input for either a formal or informal clustering procedure. There are two paths that such an engine can follow.

> Pathway 1. The user specifies (i) a point and (ii) a distance. The algorithm then (iii)

Figure 21.1. A map drawn by John Snow in 1854. The dots represent where a cholera death occurred. The crosses show the location of water pumps. Snow noticed that most of the deaths were centered on the Broad Street pump. Princeton University Library, Department of Rare Books and Special Collections.

identifies all points within that distance of that point. It is easy to imagine how a user might smoothly vary the distance and watch to see how that affects which points are identified. Such a display is dynamic in that it is the variation in identified points with changes in both the distance and the location of the chosen point that informs the viewer. Pathway 1 is especially suitable for large data sets because the calculations required are of the same order of magnitude as the number of points in the data set.

Pathway 2. The user specifies only a distance, and the algorithm then identifies all groupings of points that fall within that specified distance of one another.* Such a procedure may become computationally heavy with large data sets because the number of calculations it requires is related to the square of the number of data points.

* One such procedure is very closely akin to hierarchical clustering (Johnson 1967).

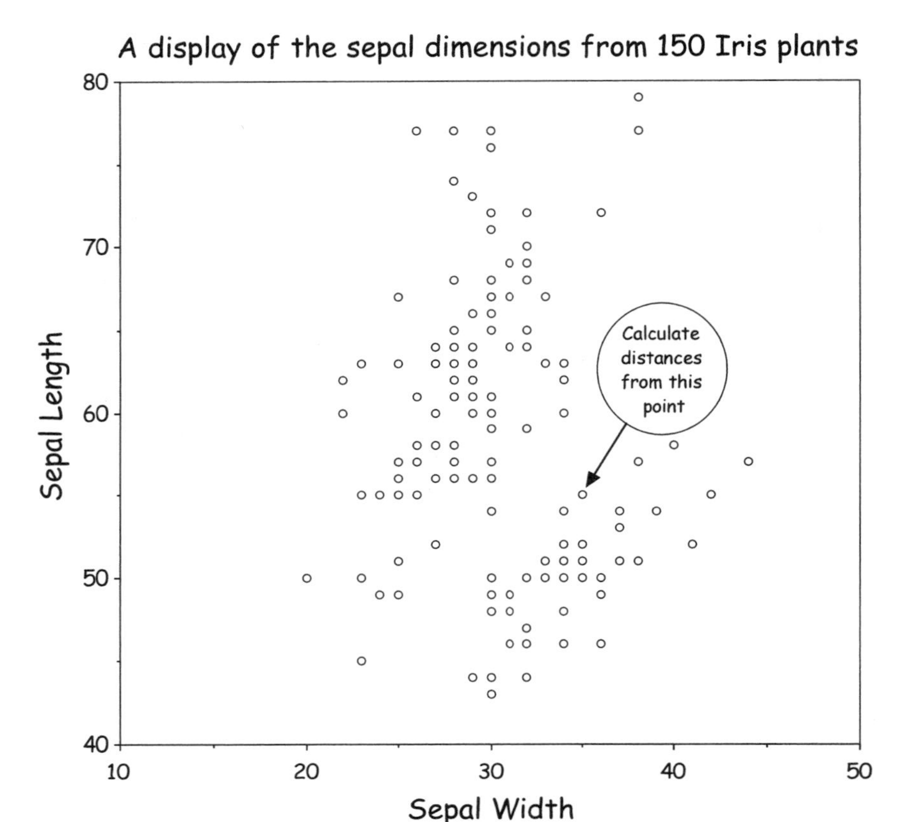

Figure 21.2. A scatterplot of sepal lengths versus widths for 150 iris plants. These measurements were drawn from fifty of each of three varieties of iris: *Iris setosa*, *Iris versicolor*, and *Iris virginica*.

Although a nearness engine can be helpful with a two-dimensional data set, its value is limited, since we can easily see how things group simply using our eyes and an intuitive sense of what "close together" means. It is of much greater value as a tool in exploring data sets of high dimensionality. In such a situation, we may be looking at a two-dimensional plot of a three-dimensional structure. Two points may look close together, but if we could but see the third dimension, we would realize they are very far apart indeed. An obvious example of this occurs when we look at the night sky and see two stars that appear very close to one another. We know that in the unseen third dimension, they may be millions of light years apart. A nearness engine that operates within the full dimensionality of the data space can be invaluable if it prevents us from making false inferences.

For a simple version, let us consider how we might apply a nearness engine to a high-dimensional rotating plot. Suppose we operationalize a nearness tool as a movable round icon the diameter of which is the diameter of the clusters we are searching for. Such a tool might be called a "brush." As we brush around the space, we are bound to find unexpected points lighting up or failing to light. These are points that look nearby or far from the center of the plot, but unexpectedly either do or do not make it inside the radius of the brush. We

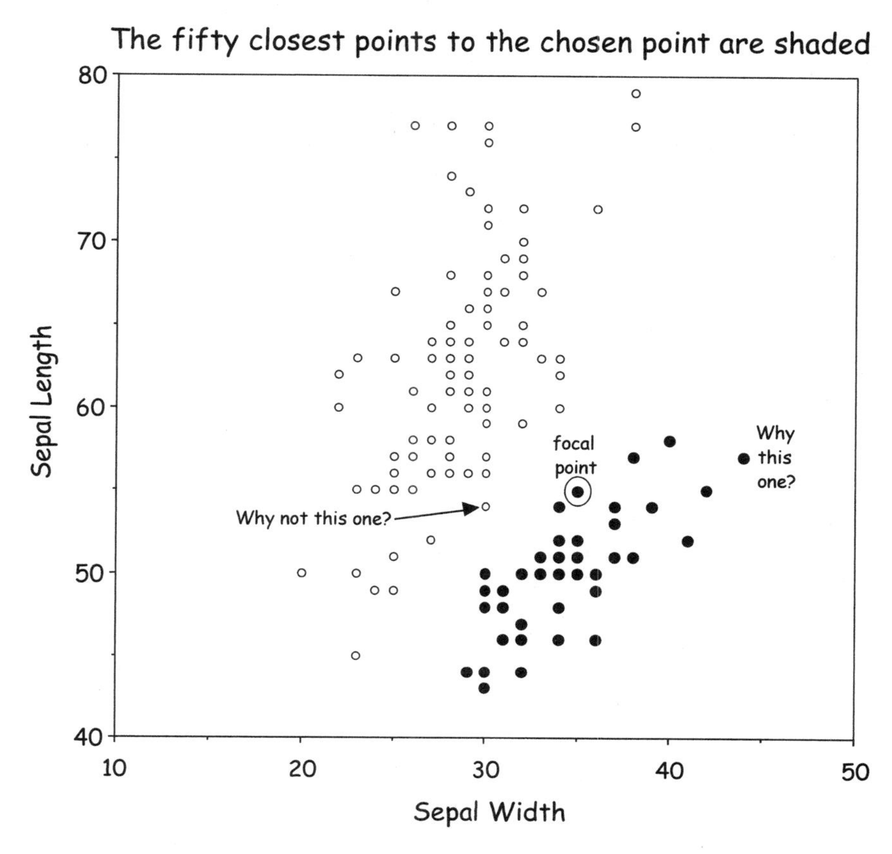

Figure 21.3. The plot in figure 21.2 with the fifty points closest to the focal point shaded. The distances were calculated in the four-dimensional space that also includes measurements of their petals.

can then leave those points highlighted and rotate to understand what has happened.*

As an example, consider the 150 data points in figure 21.2 (measurements made by the botanist Edgar Anderson, but first published in 1936 by the statistician Ronald A. Fisher). There were four measurements (in centimeters) made on each of three varieties of iris; sepal length, sepal width, petal length, and petal width. Originally, there were only two varieties, *Iris setosa* and *Iris versicolor*, but Fisher added data† on *Iris virginica* to test Randolph's hypothesis[2] that *Iris versicolor* is a polyploid hybrid of the other two species.

Of the four variables we plot two, sepal width versus sepal length, and ask our nearness engine to identify points within some distance d of the indicated point. As we increase the size of d, we find that one group of points is darkened, to the exclusion of many others that seem, in these two dimensions at least, to be much closer (figure 21.3).

Obviously, something must be happening in the two dimensions that we cannot see that

* Of course, to do this we would have to specify the center of the brush in multivariate space, but we could constrain it to a plane through the rotating plot origin (perhaps the medians of the variables) for those dimensions currently shown perpendicular to the screen in a rotating plot.

† Also gathered by Anderson (1928).

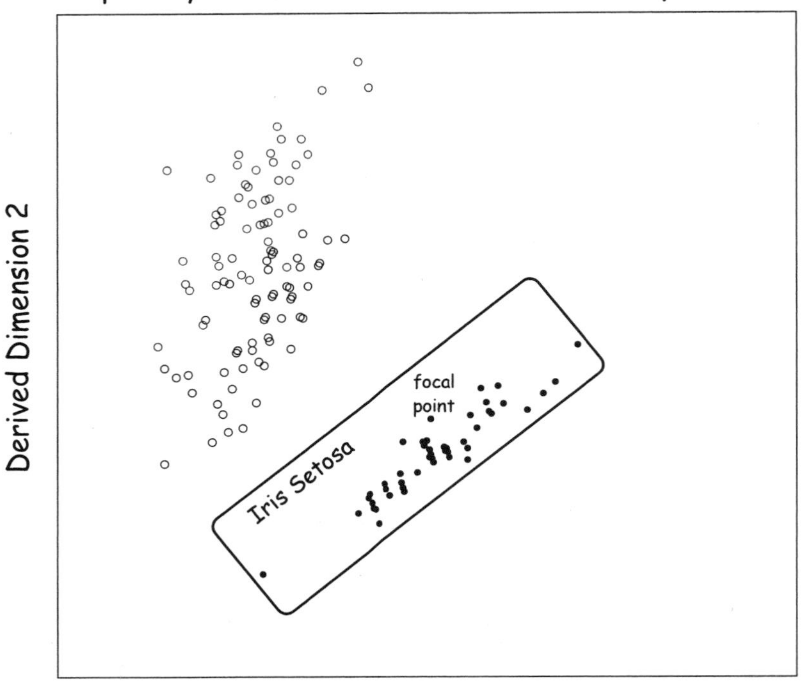

Figure 21.4. The iris data were rotated in three-dimensional space to yield an orientation in which the two-dimensional projection shown isolates the shaded points. These were all *Iris setosa* plants.

makes the identified points compact in some Euclidean way. To look for this, I plotted the points in three dimensions (adding the two petal dimensions and removing one sepal dimension) and rotated around until the identified points were maximally isolated. This result is shown in figure 21.4.

I discovered that the selected points were forty-nine of the fifty *Iris setosa* in the sample.

Tool 4. Smoothing Engine

Data points usually contain a mixture of signal and noise. Our perceptions are often aided if there is a way to filter some of the noise and present a purer visual version of the signal. One way to do this has traditionally been smoothing, and many effective smoothers have been developed. A smoother typically functions in a bivariate sense by having one coordinate carry a value and a second coordinate carry the order. The smoother modifies the values on the basis of the order. One important advantage of using general smoothers, rather than a more traditional sort of functional fit, is that smoothers are local. That is, the effects of some unusual points on a smoothed fit affect only those parts of the fit near those points. This is not true in general for traditional fit functions (e.g., polynomials), in which data points have a more global effect.

As an example, consider once again the now familiar data shown in figure 21.5, which plots all of the winning times in the Boston Marathon for men and women. A linear fit to each set of points suggests that starting in the year 2005, women will begin to run the marathon faster than men. Quite a different interpretation is yielded from the smoothed fits shown to the same data in figure 21.6.*

Smoothing when there is only one order dimension and one value dimension is easy. When there are several of each it is hard.

Concluding Remarks

The content of this chapter and the one that preceded it came directly from conversations that I had with John Tukey 1999 and 2000 on this topic. The immediate goal of these discussions was to have something worthwhile to say on the future of graphics at a conference in Chicago commemorating the anniversary of the publication of Tukey's pathbreaking book, *Exploratory Data Analysis*. The conference was held on May 5, 2000, but because of his precarious health John decided against attending, although he did put in a telephonic appearance and answered questions.

After this conference, we continued to discuss and refine the concepts, and even worked on it some more during breaks in the eighty-fifth birthday shindig for him that was thrown by the New Jersey chapter of the American Statistical Association on June 16, 2000. The last time we spoke about this topic was over mince pie (his favorite) in his hospital room on July 24, 2000. When I left him that evening, he said that he was tolerably happy with where we were, at least for the moment, but we would revisit it when we next got together. Unfortunately, this was as far as we got, for John Wilder Tukey died of an accumulation of physical difficulties early in the morning of July 26th.

That John would choose to spend the last few hours of his life working on statistical problems surprises no one who knew him. Shortly after John's death, the Stanford statistician and McArthur Prize recipient David Donoho (John supervised David's senior thesis at Princeton) wrote,

> After John got an honorary degree from Princeton last year, he called me about organizing a project with some astronomers to develop some new kinds of signal processing and analysis. This possibility had arisen when he was chatting with the provost at the degree ceremony!
>
> What incredible intellectual vitality he had—seeking out new scientific vistas in the middle of a ceremony where 99.999 percent of the people are just happy to be getting their certificates of accomplishment. (Personal communication, August 2000)

At his wife Elizabeth's funeral, two years previous, I had mentioned to John that I would like to call on him in a couple of weeks (or whenever he was feeling up to it) to discuss a statistical problem I had. John asked, "What about now?" I told him it could wait, but he insisted. So we sat there, amidst vari-

* This particular smooth is 53h, described by Tukey (1977). It involves first taking medians every 5 points. This yields considerable smoothing. Next resmooth by smoothing those medians every 3. The h is a specific linear combination of the 3. The resulting smoothed vector can then be subtracted from the original data and the process repeated on the residuals, but why that is sensible is a topic for another day. 53h is a descriptive designation for many classes of smoothers—one could easily imagine using 75h or 97h for noisier data.

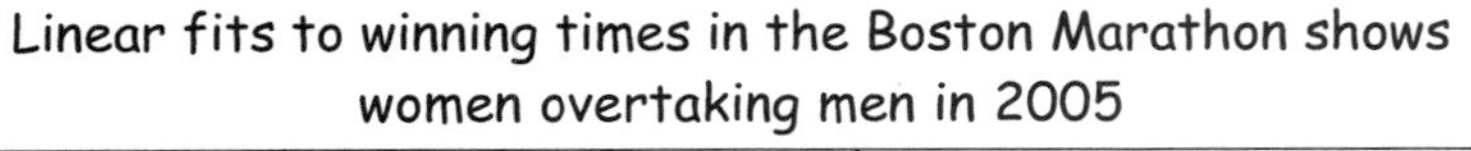

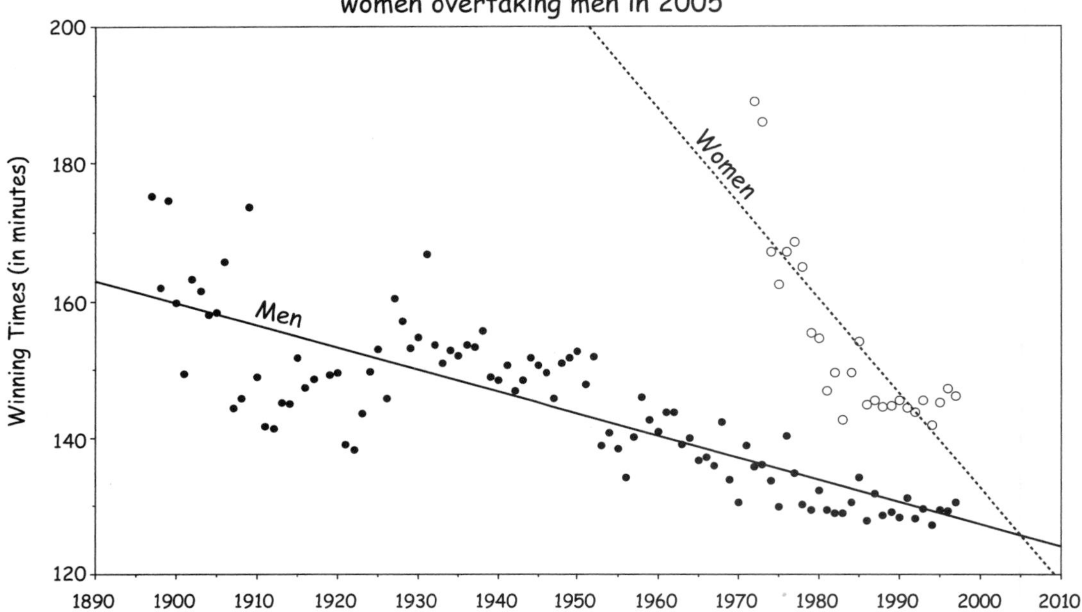

Figure 21.5. The winning times for the Boston Marathon from 1897 (for men) and 1972 (for women) until 1997. The best-fitting straight lines are shown for each sex.

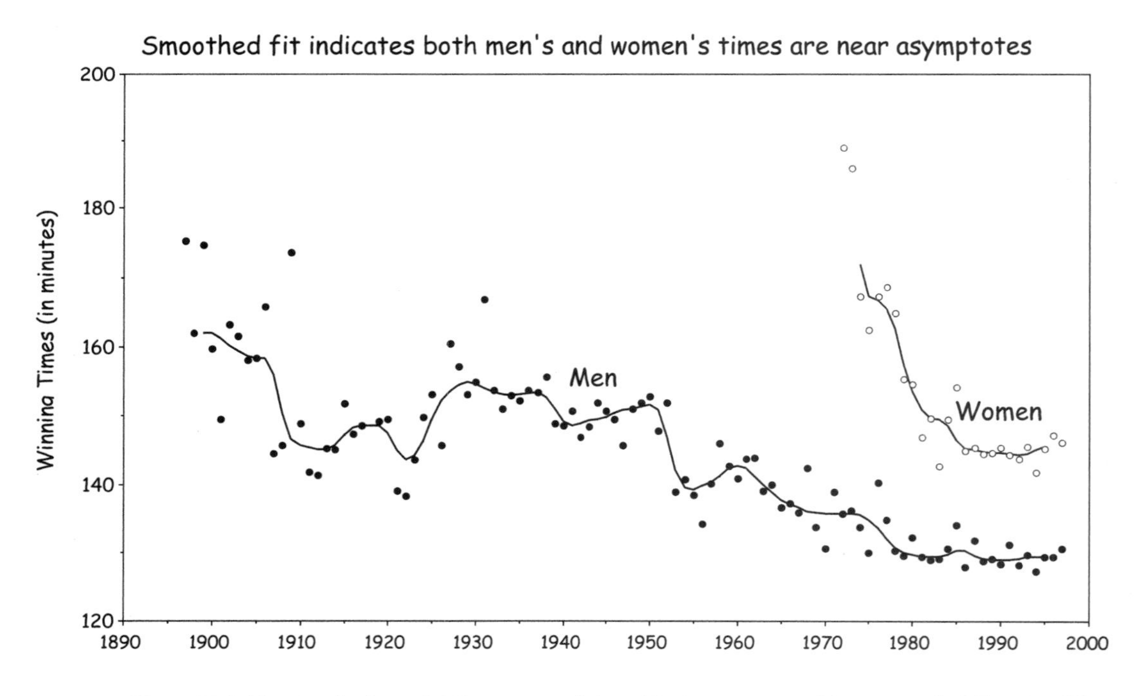

Figure 21.6. The data in figure 21.5 are repeated, but this time augmented by a smoothed function. The smoothing method used is 53h, which primarily involves calculating running medians.

ous friends and relatives, who stopped and tenderly offered condolences, and worked on a complex prediction problem.

Zorba, the famous Greek philosopher, observed, "When my son Dimitri died, I danced. They all thought I was mad, but it was the only thing that made the pain go away." John and I danced together for most of that afternoon.

John Tukey's funeral was held on Monday, July 31, 2000, at Trinity Church in Princeton. The church was full of statisticians. I was reminded of John Kennedy's comment at a White House dinner honoring Nobel Prize winners, "I think this is the most extraordinary collection of talent, of human knowledge, that has ever been gathered together at the White House, with the possible exception of when Thomas Jefferson dined alone." I had not seen such an accumulation of statistical horsepower in one place since the previous Monday, when John sat in his hospital room and ate his mince pie.

After the funeral, a person I did not recognize came up to me and asked,

"You have been working with Professor Tukey for many years. Have you learned his methods for solving problems?"

"Yes," I replied, "I have."

"Could you tell me?" he inquired.

"Certainly," I said, "He used a three step approach.

"First, he would listen very carefully to the problem, asking questions when necessary to clarify the issues in his own mind.

"Second, he would sit quietly for however long it took, and think very hard.

"Then, third, he would give you the answer."

What appears in these two chapters is the direct result of his following this three-step approach.

22 Epilogue: A Selection of Selection Anomalies

> The government [is] extremely fond of amassing great quantities of statistics. These are raised to the nth degree, the cube roots are extracted, and the results are arranged into elaborate and impressive displays. What must be kept ever in mind, however, is that in every case, the figures are first put down by a village watchman, and he puts down anything he damn well pleases.
>
> —Sir Josiah Charles Stamp, 1880–1941

The focus of this book has been on displaying data and, except in chapter 10, I have spent very little effort discussing the quality of what is being displayed. In good conscience I cannot end without at least providing a modest caveat. The consideration of the quality of the data that we plot is a side trip of this particular voyage, so I have opted to use the space necessary for only a few short stories to illustrate the often subtle concerns that can yield potentially profound consequences. And last, I am unable to resist the opportunity to preach a little about the character of a solution to self-selection, perhaps the most pernicious obstacle to making accurate inferences.

Every good teacher knows that the instructor's initial task must be to motivate the students. Motivation is an especially daunting task for those in my field, for it is a rare student indeed that looks forward to the study of statistics. Implicit in the sea of glazed faces that typically greet me on the first day of class is the question, "Why do I need to know this stuff?" The answer, which is never believed without proof, is "To be able to understand the world." Generating a proof of this statement requires a convincing, real example. I am happy to provide one.

Most students can easily concoct a situation in which it is important to know how many children are in the average family. But how would we find out the answer? When this question is posed, several blasé hands inevitably go up or someone shouts out, "Do a survey." "Good," I respond, drawing them into my net, "what do we ask in the survey?" The happy reply emerges quickly, "How many children were in your family?" When I get this response, I immediately ask everyone in the class to respond to this question. We then collect the results, add them up and calculate the average. This is never very far from 3.5. Inwardly smirking, I say, "The usual figure that the Census Bureau gets is about 2.5. Why are we so far off?" At this point, I am treated to explanations of random fluctuations, small samples, old data, governmental incompetence, and bad luck.

But are those indeed the answer? I ask, "Do any of you know of any families of your parents' generation that had no children?" Clearly, such families would have no representatives in my class. In the same way, a family that had six children would have six times the chance to be represented than a family with only one child. Slowly the answer dawns: the bigger the family, the greater the chance that they will be represented in the sample. This yields a response bias toward a larger-than-correct answer.* If getting the right answer to such a simple question requires subtle thinking, how can we learn the answers to more difficult questions, such as "Does mammography reduce the likelihood of death from breast cancer?" To answer such questions one must know a considerable amount about statistical thinking, for in this arena at least, a little ignorance is a dangerous thing.

Gathering data, like making love, is one of those activities that almost everyone thinks can be done well without instruction. The results are usually disastrous. Let me elaborate.

During the 1984 U.S. presidential campaign, the Democratic vice-presidential nominee, Geraldine Ferraro, was asked how she could persevere in the face of very discouraging poll results. She said, "I don't believe those polls. If you could see the enthusiasm for our candidacy out there, you wouldn't believe them either." Of course, part of her response must have been political hyperbole, but even after the election, when the polls' predictions proved to be accurate, she remained dismayed by the results. Why? It was difficult for Ms. Ferraro to get an accurate reading of her popularity in the general population from the enthusiasm she saw at Democratic gatherings. The reason for this difficulty is that the individuals who showed up at such gatherings chose to do so. They were a long way from a random sample of the population of voters. Errors of inference obtained from a nonrandom sample did not originate with Geraldine Ferraro. The telephone poll taken by the *Literary Digest* in 1936 predicting Alf Landon's victory over Roosevelt is a well-known precursor; at that time, many more Republican voters than Democrats had phones.

I would like to illustrate three circumstances where nonrandom samples could lead investigators far astray. These investigators range from a nineteenth-century Swiss physician to modern educational theorists. John Tukey used to say that there were two kinds of lawyers: one kind tells you that you can't do it, and the other kind tells you how to do it. To avoid being the wrong kind of statistician, the fourth example I present illustrates one way to draw correct inferences from a nonrandom sample with Abraham Wald's ingenious model for aircraft armoring.

Example 1. The Most Dangerous Profession

In 1835, the Swiss physician H. C. Lombard published the results of a study on the longevity of various professions. His data were very extensive, consisting of death certificates gathered over more than a half-century in Geneva. Each certificate contained the name of the deceased, the deceased's profession, and age at death. Lombard used these data to calculate the mean longevity associated with each profession. Lombard's methodology was not original with him, but instead was merely an extension of a study carried out by R. R. Madden, Esq., that had been published two years earlier. Lombard found that the average age of death for the various professions mostly ranged from the early fifties to the mid sixties.

* Note that this bias disappears if you instead ask, "How many children do you have?" But even with this question, it is still important to have a sample of respondents who are, in all relevant ways, representative of the country as a whole.

These were somewhat younger than those found by Madden, but this was expected since Lombard was dealing with ordinary people rather than the "geniuses" in Madden's study (the positive correlation between fame and longevity was well known even then). But Lombard's study yielded one surprise: the most dangerous profession—the one with the shortest longevity—was that of "student," with an average age at death of only 20.7! Lombard recognized the reason for this anomaly but apparently did not connect it to his other results.

Example 2. The Twentieth Century Was a Dangerous Time

In 1997, to revisit Lombard's methodology, my son Sam gathered 204 birth and death dates from the Princeton, New Jersey, Cemetery.[1] This cemetery opened in the mid-1700s, and has people buried in it born in the early part of that century. Those interred include Grover Cleveland, John von Neumann, Kurt Gödel, and John Tukey.

When age at death was plotted as a function of birth year (after suitable smoothing to make the picture coherent) we see the result shown as figure 22.1.* We find that the age of death stays relatively constant until 1920, when the longevity of the people in the cemetery begins to decline rapidly. The average age of death decreases from around seventy years of age in the 1900s to as low as ten in the 1980s. It becomes obvious immediately that there must be a reason for the anomaly in the data (what we might call the "Lombard Surprise"), but what? Was it a war or a plague that caused the rapid decline? Has a neonatal section been added to the cemetery? Was it opened to poor people only after 1920? Obviously, the reason for the decline is nonrandom sampling: people cannot be buried in the cemetery if they are not already dead. Relatively few people born in the 1980s are buried in the cemetery, and thus no one born in the 1980s that we found in Princeton Cemetery could have been older than seventeen.

There are many examples of situations in which this anomaly arises. Four of these are:

(1) In one hundred autopsies, a significant relationship was found between age at death and the length of the life-line on the palm.[2] Actually, what they discovered was that wrinkles and old age go together.

(2) In 90 percent of all deaths resulting from barroom brawls, the victim was the one who instigated the fight. One questions the wit of the remaining 10 percent who did not point at the body on floor when the police asked, "Who started this?"

(3) In March 1991, the *New York Times* reported the results of a poll by the American Society of Podiatry, which stated that 88 percent of all women wear shoes at least one size too small. Who would be most likely to participate in such a poll?

(4) In testimony before a February 1992 meeting of a committee of the Hawaii state senate, then considering a law requiring all motorcyclists to wear helmets, one witness declared that despite his having been in several accidents during his twenty years of motorcycle riding, a helmet would not have prevented any of the injuries he received. Who was unable to testify? Why?

* The points shown in figure 22.1 are those points generated by a nonlinear smoothing of the raw data. The smoother used was 53h-twice (Tukey 1977), which involves taking running medians every 5, then running medians of those every 3, then a weighted linear combination. This smoothed estimate is then subtracted from the original data and the process repeated on the residuals. The resulting two smooths are then added together for the final estimate shown.

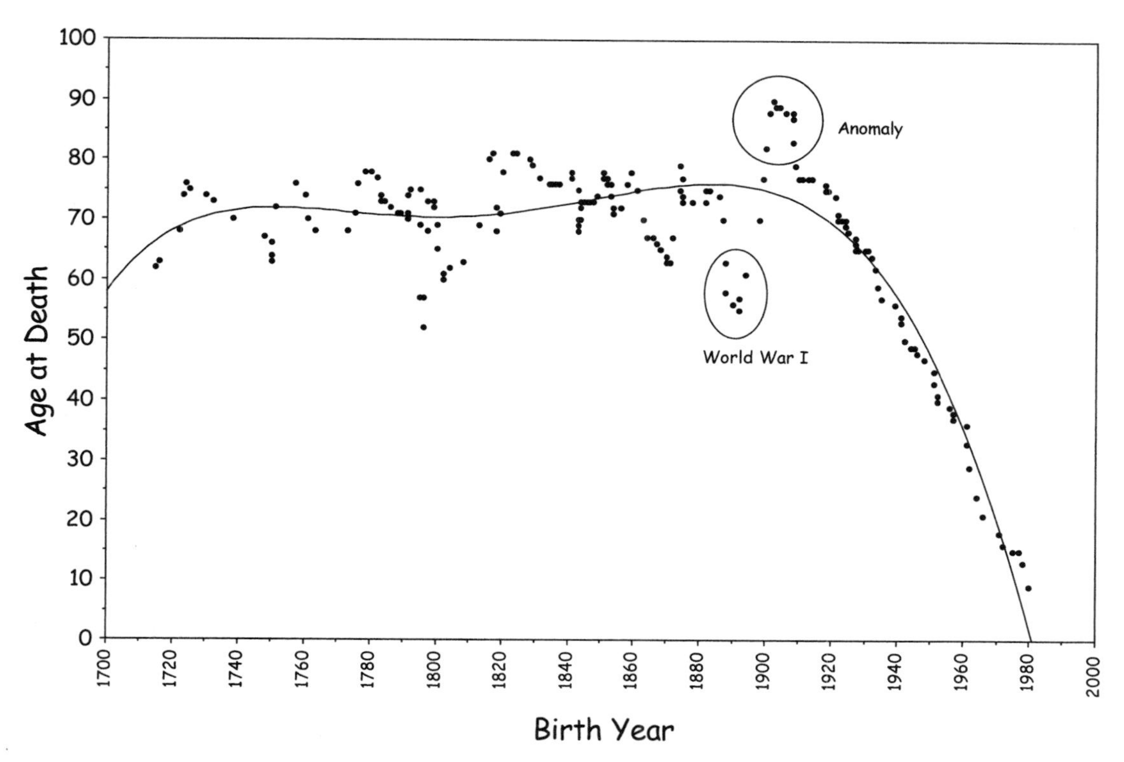

Figure 22.1. The longevities of 204 people buried in Princeton Cemetery shown as a function of the year of their birth. The data points were smoothed using "53h twice" (Tukey 1977), an iterative procedure of running medians.

Selection, or, more generally, nonrandom sampling, is often as subtle in its manifestation as it is substantial in its effect. I have so far emphasized the size of its effects. Next let us consider an instance of selection in education whose interpretation has yielded substantial debate among experts.

Example 3. What Do Changing SAT Scores Mean?

The Scholastic Assessment Test (SAT) is taken by more than a million high school seniors annually. Since 1962, the average SAT score has declined (see figure 22.2). Many have interpreted this decline as an indicator of the failure of the American educational system, although its principal causes remain in question.

A national panel created by the College Board in 1977 placed most of the blame on students taking too many watered-down courses.

Gene Maeroff, in a front-page story in the *New York Times* (September 22, 1982), suggested that the test results were further "evidence that a loosening of high school requirements in the 1960s had led to a deterioration of educational standards." Maeroff also cited a 1980 article in the *Phi Delta Kappan*, a respected educational journal, that "attributed the score decline to the effect on the young of fallout from above-ground nuclear tests of the 1950's and early 1960's."

Time magazine (October 4, 1982) suggested that a mix of "social factors, including television, the frequency of divorce and the

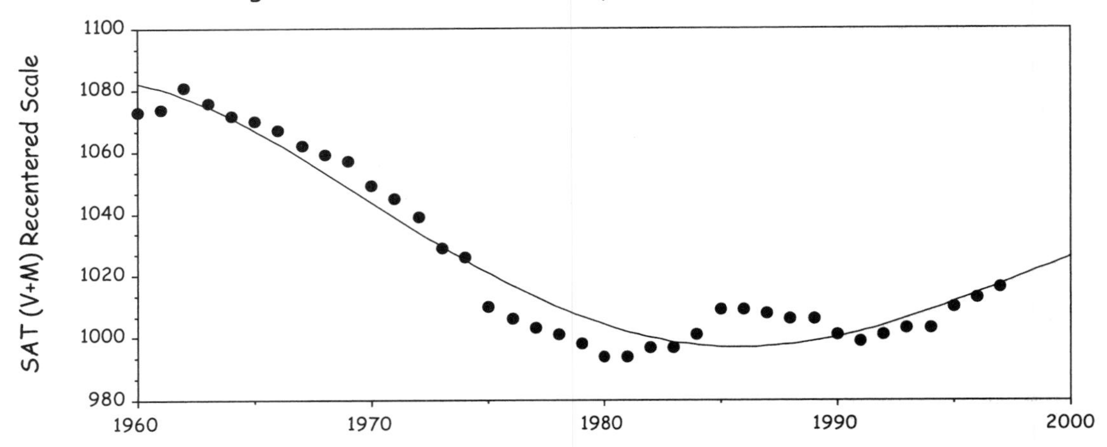

Figure 22.2. The mean SAT score (verbal plus mathematical) for all high school seniors in the United States who took the test since 1960. These scores are shown on the (April 1996) recentered scale.

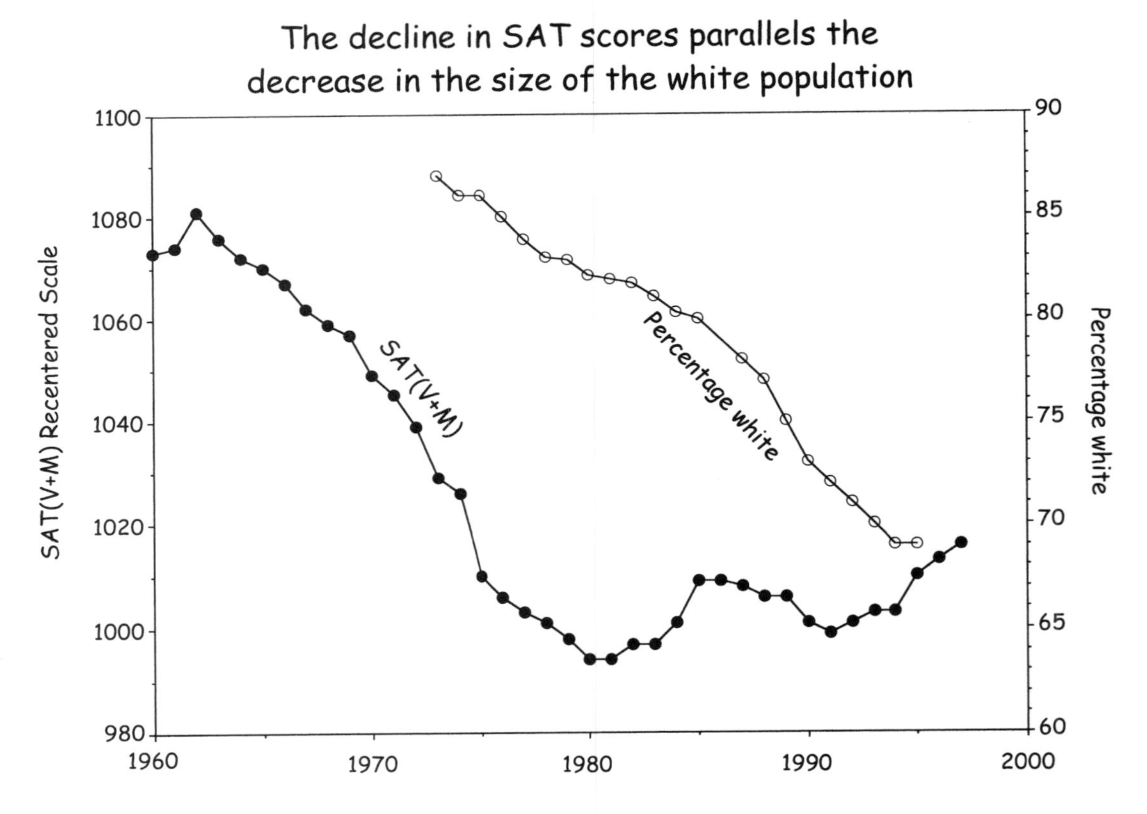

Figure 22.3. The mean SAT score shown on the same graph as the percentage of the U.S. population that classified themselves as "white" on the Current Population Survey. The decline of the SAT parallels the decline of the white population.

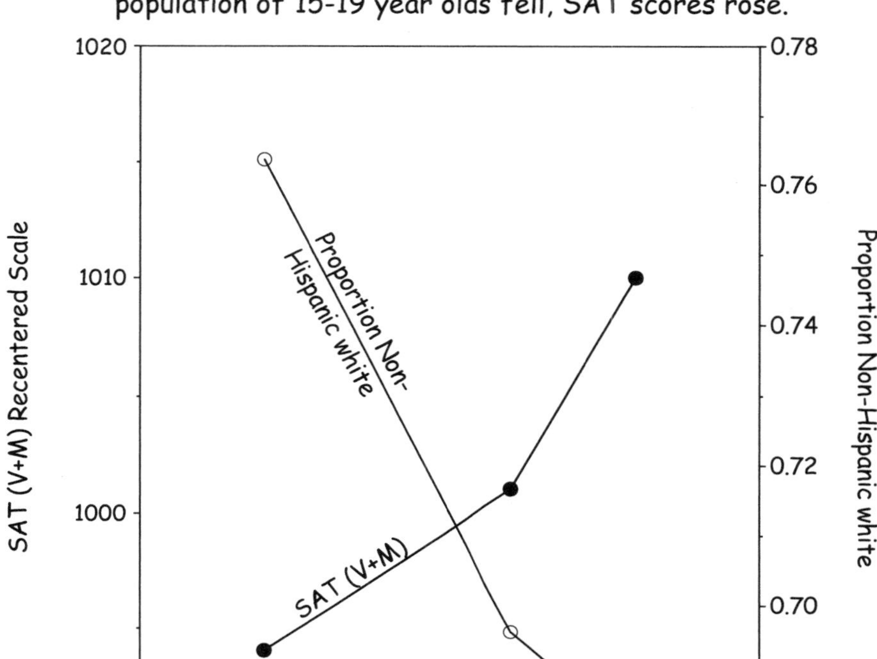

Figure 22.4. The mean SAT score for 1980–1995 shown on the same graph as the proportion of the U.S. population, ages 15–19, that classified themselves as "white" and "non-Hispanic" on the Current Population Survey. The increase of the mean SAT score stands in stark contrast to the decline of the white population.

softening of high school curriculums," caused the score decline.

Laura Durkin, in *Newsday* (September 22, 1982), understood the effects of selection and attributed the decline to "the much larger pool of students taking the test in recent years."

In the early to mid-1980s, a consensus began to emerge to explain this trend. The essence of this argument was that an increasing proportion of students in the senior class around the nation took the SAT and that this group included many minority students, who historically have not done well on standardized tests. This opinion was borne out by such evidence as that shown in figure 22.3, in which the declines in SAT scores are paralleled by declines in the proportion of the population that is white.

This interpretation falls flat when data since 1985 are examined, which show exactly the opposite result (see figure 22.4).

How can we draw valid inferences from nonrandomly sampled data? The answer is "not easily" and certainly not without risk. The only way to draw inferences is if we have a model for the mechanism by which the data were sampled. Let us consider one well-known example of such a model.

Example 4. Bullet Holes and a Model for Missing Data

Abraham Wald, in some work he did during World War II,[3] was trying to determine where to add extra armor to planes on the basis of the pattern of bullet holes in returning aircraft. His conclusion was to determine carefully where returning planes had been shot and *put extra armor every place else*!

Wald made his discovery by drawing an outline of a plane (crudely shown in figure 22.5) and then putting a mark on it where a returning aircraft had been shot. Soon the entire plane had been covered with marks *except* for a few key areas. It was at this point that he interposed a model for the missing data, the planes that did not return. He assumed that planes had been hit more or less uniformly, and hence those aircraft hit in the unmarked places had been unable to return; thus the unmarked places were the areas that required more armor.

Wald's key insight was his model for the nonresponse. From his observation that planes hit in certain areas were still able to return to base, Wald inferred that the planes that did not return must have been hit somewhere else. Note that if he used a different model, analogous to "those lying within Princeton Cemetery have the same longevity as those without" (i.e., that the planes that returned were hit about the same as those that did not return), he would have arrived at exactly the opposite (and wrong) conclusion.

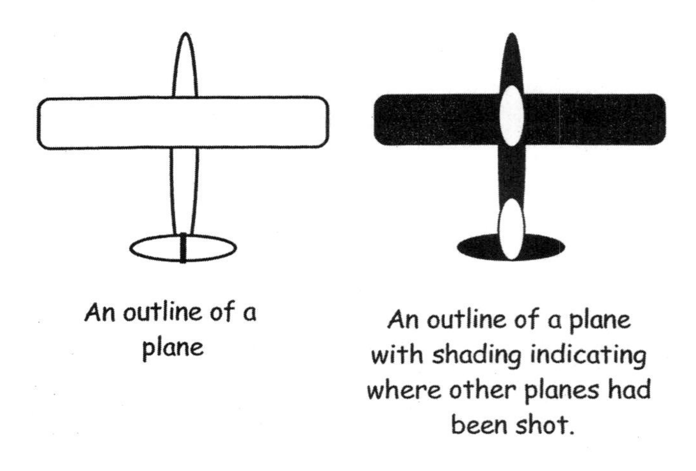

Figure 22.5. A schematic representation of Abraham Wald's ingenious method to investigate where to armor aircraft.

To test Wald's model requires heroic efforts. Planes that did not return must be found and the patterns of bullet holes in them must be recorded. In short, to test the validity of Wald's model for missing data requires that we sample from the unselected population—we must try to get a random sample, even if it is a small one. This strategy remains the basis for the only empirical solution to making inferences from nonrandom samples.

In the cemetery example, if we want to get an unbiased estimate of longevities we might halt our data series at a birth date of 1900. For the SAT, a number of selection models have been tried,[4] and all have failed to be accurate enough for the purposes they were intended. We are thus drawn inexorably toward the conclusion that the best way to get an accurate indicator of average student performance is through a survey constituted from a well-designed, rational sampling process.

A Concluding Parable

I do not mean to suggest that it is impossible to gain useful insights from nonrandomly selected data, only that it is difficult and great care must be taken in drawing inferences. James Thurber's (1939) *Fables for Our Time* tells the story of "The Glass in the Field." It seems that a builder left a huge pane of window glass standing upright in a field one day. Flying at high speed, a goldfinch hit the glass and was struck senseless. Later, upon recovering his wits, he told a seagull, a hawk, an eagle, and a swallow about his injuries caused by crystallized air. The gull, the hawk, and the eagle laughed and bet the goldfinch a dozen worms that they could fly the same route without encountering crystallized air, but the swallow declined and was alone in escaping injury. Thurber's moral: "He who hesitates is sometimes saved." This is my main point: that a degree of safety can exist when one makes inferences from nonrandomly selected data, if those inferences are made with caution. There are some simple methods available that help us draw inferences when caution is warranted; they ought to be used.

This book is an inappropriate vehicle to discuss these special methods for inference in detail.[5] Instead, let us describe the general character of any "solution." First, no one should be deluded into thinking that when there is a nonrandom sample, unambiguous inferences can be made. They cannot. The magic of statistics cannot create information when there is none. We cannot know for sure the longevity of those who are still alive or the SAT scores for those who did not take the test. Any inferences that involve such information are doomed to be equivocal. What can we do? One approach is to make up data that might plausibly have come from the unsampled population (i.e., from some hypothesized selection model) and include them with our sample as if they were real. Then see what inferences we would draw. Next, make up some other data and see what inferences are suggested. Continue making up data until all plausible possibilities are covered. When this is done, see how stable were the inferences drawn over the entire range of these data imputations. The multiple imputations may not give a good answer, but they can provide an estimate of how sensitive inferences are to the unknown. If you do not do this, you have not dealt with possible selection biases, you have only ignored them.

Conclusion

In 1923, Edna St. Vincent Millay, applauding the growth of good data while simultaneously decrying the lack of good techniques for exploratory data analysis, wrote:

> Upon this gifted age in its dark hour
> Falls from the sky a meteoric shower
> Of facts. They lie, unquestioned, uncombined.
> Wisdom enough to leach us of our ills is daily spun,
> But there exists no loom to weave it into fabric.

This book celebrates more than two centuries of progress toward the construction of a glorious loom.

Dramatis Personae

This centuries-long tale of the visual display of quantitative phenomena has a very large cast. To aid the reader in keeping track of who was who I have included brief biographical sketches of all but the bit players. I hope this is as much fun to read as it was to prepare.

Albee, Edward (1928–) American playwright born in the Washington, DC, area and educated at Trinity College and Columbia University. Among his major works are *Zoo Story* (1958), *Who's Afraid of Virginia Woolf?* (1962), and *A Delicate Balance* (1966). He won a Pulitzer Prize for *Three Tall Women* (1991).

Anderson, Edgar (1897–1969) Evolutionary botanist, biosystematist, and author of *Plants, Man and Life* and *Introgressive Hybridization*, who spent the bulk of his career at the Missouri Botanical Garden. Intensely interested in developing graphical techniques for capturing variation in natural populations, he worked with Sewall Wright, R. A. Fisher, and John Tukey. His iris data were used by Fisher for the 1936 presentation of the linear discriminant function. With Tukey he developed his pictorialized scatter diagrams into the more dynamic metroglyphs. Tukey's *Exploratory Data Analysis* is dedicated, in part, to Anderson.

Baker, Stuart G. (1958–) Born in Washington, DC, son of an architect. After studying applied mathematics and biostatistics at Harvard, he went to work at the National Cancer Institute. He has done fundamental work on missing categorical data, causal inference (the paired availability design for historical controls), and evaluating biomarkers and screening for the early detection of cancer.

Barbeau-Dubourg, Jacques (1709–1779) A French *esprit éclairé*, he successively learned theology, law, and medicine, demonstrating a breadth of scholarship and accomplishment that remains astonishing to this day. He was born in Paris and graduated from the Paris Medical School in 1748. In addition to inventing the chronological machine described in chapter 7, he was the editor of the medical journals *La Gazette d'Épidaure* and *La Gazette de médecine*. As a practicing physician, he was a strong advocate of inoculation to prevent the variola epidemic. He was also an accomplished botanist, publishing *Le Botaniste françois, Manuel d'herborisation, Les Âges des plantes*, and *Manuel de botanique*. But such activities did not prevent him from pursuing his philosophic interests and publishing the moralist tract *Petit code de la raison humaine*. He numbered many scientists among his friends and correspondents, with a special affection for Benjamin Franklin and America. In his career as a pamphleteer, he published a number of pro-American documents that included a French translation of his extensive correspondence with Franklin as well as *Lettres d'un fermier de Pennsylvanie aux habitants d'Amérique septentrionale, Calendrier de Philadelphie.*

Barère (de Vieuzac), Bertrand (1755–1841) Revolutionary and regicide born in Tarbes, France, who, though originally a moderate in the National Convention, helped form the Committee of Public Safety. He was imprisoned after the fall of Robespierre (1794) but escaped. He later served under Napoleon and was exiled with him, only to return to Paris under an amnesty in 1830.

Bayes, Thomas (1702–1761) Born in London, he was one of the first six Nonconformist ministers to be publicly ordained in England. He is principally remembered for his posthumously published "Essay towards Solving a Problem in the Doctrine of Chances" (1763), in which he proved a theorem, since named for him, that allows the representation of a conditional probability in terms of its reverse: $P(A|B) = P(B|A)\ P(A)/P(B)$. This theorem has generated an entire school of statistical thought, whose members, naturally enough, call themselves Bayesians.

Bernoulli, Daniel (1700–1782) A scientific generalist who was born in Groningen, The Netherlands, son of Johann Bernoulli. He studied medicine and mathematics. He began his professional career as a professor of mathematics at St. Petersburg in 1725. In 1732 he went to Basel to become professor of anatomy, then botany, and finally physics. He solved a differential equation proposed by Jacopo Riccati, now known as Bernoulli's equation, and was a pioneer in the development of utility theory.

Berra, Lawrence Peter ("Yogi") (1925–) American-born pundit and catcher. He played with the New York Yankees from 1946 until 1963 and during that time played in fourteen World Series (a record). He set the record for the most home runs by an American League catcher (313). After his playing days were over, he managed the Yankees, the Mets, and the Astros. He is famous for a plethora of Yogi-isms. Among the best known are: "it ain't over 'til it's over," "they pay you in cash, which is just as good as money," and "nobody goes there anymore 'cause it's too crowded."

Bertin, Jacques (1916–) French semiologist, trained in Paris, whose seminal work *La Semiologie graphique* (1969) laid the groundwork for modern research in graphics. Until his retirement in 1984, he was professor at the Laboratoire graphique in the École des Hautes Études in Paris.

Biderman, Albert (1922–) American sociologist trained at the University of Chicago. He was instrumental in founding the National Science Foundation (NSF) working group on social graphics, and his NSF project on graphics can be credited with instigating the graphics work of Edward Tufte and Howard Wainer.

Bok, Derek (1930–) American lawyer and former president of Harvard University. He was born in Bryn Mawr, Pennsylvania, and educated at Stanford and Harvard Law School. He moved from professor of law at Harvard (1958) to dean of the law school (1968) to president of the university (1971–1991). He is the author of *Labor and the American Community* (1970), *Beyond the Ivory Tower* (1984), *The State of the Nation* (1997), and *The Trouble with Government* (2001). He is currently a professor at Harvard as well as director of the Hauser Center for Nonprofit Organizations.

Bowen, William (1928–) American economist and educator. He was educated at Princeton, where, after he finished his Ph.D. (1958), he joined the faculty of economics. Following a research trip to England he published *Economic Aspects of Education* (1964), which was quickly followed by *Performing Arts: The Economic Dilemma* (with William Baumol, 1966), and then *The Economics of Labor Force Participation* (with T. A. Finegan, 1969). At this point he became Princeton's provost under President Robert Goheen and most of his scholarship ceased. When Goheen retired in June 1972, Bowen became Princeton's president, and remained so for fifteen years. He is currently president of the Andrew W. Mellon Foundation.

Brahe, Tycho (1546–1601) Swedish astronomer. In 1573, at age seventeen, he discovered serious errors in the astronomical tables and began work to correct this by observing the stars and planets with unprecedented accuracy. Ironically, although he rejected the Copernican theory, it was his data that Kepler used to confirm it.

Breyer, Stephen (1938–) American jurist born in San Francisco and educated at Stanford, Oxford, and Harvard Law School (1964). After clerking for Arthur Goldberg (1964–1965), he taught at Harvard until 1980, when he was named to the First Circuit of the U.S. Court of Appeals by Jimmy Carter. He was appointed as an associate justice to the Supreme Court of the United States in 1994 by Bill Clinton. Among the books he has published are *Regulation and Its Reform* (1982) and *Breaking the Vicious Circle: Toward Effective Risk Regulation* (1993).

Brinton, Willard C. (1880–1957) American construction engineer born in West Chester, Pennsylvania, and educated at Harvard (1907). He had an extensive career as an engineer working for a variety of companies before he began his own consulting company situated in New York City. He was instrumental in the establishment of the Port Authority of New York. He slid into statistics, and by 1919 he had been elected vice-president and director of the American Statistical Association. He authored *Graphic Methods for Presenting Facts* (1914) and *Graphic Presentation* (1939).

Bronowski, Jacob (1908–1974) Mathematician, poet, and humanist. He was born in Lodz, Poland, the son of a haberdasher who emigrated to England in

1920. He studied mathematics at Cambridge, focusing on algebraic geometry and topology, and completed his Ph.D. in 1933. While his interests in mathematics were subsequently extended to statistics and mathematical biology, his friendships with Laura Riding and Robert Graves stimulated his interest in poetry. His major works reflect the breadth of his interests. His first major book, *The Poet's Defence* (1939), examined the relationship between science and poetry; this was followed by *William Blake, a Man without a Mask* (1944). As part of his work with the British Mission to Japan after World War II, he wrote *Effects of the Atomic Bombs at Hiroshima and Nagasaki* (1947). In 1953, while at MIT he developed a set of lectures that became *Science and Human Values* (1961). Later, while at the Salk Institute for Biological Studies he filmed a thirteen-part television series for the BBC, *The Ascent of Man* (1971–1972), which later appeared in book form.

Burt, Sir Cyril (1883–1971) British psychologist trained at Oxford and Würzburg. He was a professor of education (1924–1931) and psychology (1931–1950) at University College, London, and was knighted in 1946. His work on intelligence testing was pathbreaking, but after his death profound errors in his empirical claims were uncovered and called into question large portions of his life's work. *See* Jane Conway.

Carnevale, Anthony (1945–) Former vice-president of the Educational Testing Service who devised the notion of a "striver," operationalized it, and called a press conference to popularize it. When it was pointed out that his view of this was incorrect (indeed backward), he refocused his efforts on other topics.

Celeste, Suor Maria (1600–1633) Galileo's illegitimate daughter, who lived out her life in a nunnery but through extraordinary, unselfish efforts helped to keep her father healthy and productive. Her own diet was so scant that she died of dysentery, which she probably would have survived had she been stronger.

Cleveland, William (1943–) American statistician trained at Princeton and Yale whose work on the scientific foundations of graphical perception demonstrates a unique melding of expertise in statistics and psychology. This work was done at Bell Labs and yielded a series of monographs that provide evidence-based guidance to the design of data-analytic graphics. His best-known works are *The Elements of Graphing Data* (1994), *Visualizing Data* (1993), and (with Marilyn McGill) *Dynamic Graphics for Statistics* (1988).

Condorcet, Marie Jean Antoine Nicolas de Caritat, Marquis de (1743–1794) A French philosopher, mathematician, and statesman. He was trained in Paris and his mathematical work was highly regarded in the 1760s. He then went astray, getting involved in politics in which, during the Revolution, his speeches and pamphlets allied him with the Girondists. He was selected for the Legislative Assembly (1791) and became its president. He eventually ran afoul of the extremists, who arrested him. He died in prison.

Conway, Jane An assistant to Cyril Burt. Little is known about her. She published several important empirical confirmations of Burt's theory of the heritability of intelligence in the *British Journal of Mathematical and Statistical Psychology*, which Burt edited. L. S. Hearnshaw, a posthumous biographer of Burt, who had access to his private papers, provided strong evidence that she never existed and was a pseudonym Burt used to get results published from what appeared to be an independent source.

Cromwell, Thomas, Earl of Essex (1485–1540) An English statesman who had a successful career as an administrator and advisor to the king. Among his other accomplishments, he arranged Henry VIII's divorce from Catherine of Aragon (and was largely responsible for the beheading of Sir Thomas More, 1478–1535, in the process). Ironically, five years later, he was done in by Henry's aversion to Anne of Cleves, a consort of Cromwell's choosing: having failed to please the king, he was sent to the Tower of London and beheaded. Perhaps specifying the consequences of failure in this way would provide a workable pathway toward improving the efficacy of computer dating services.

Davis, Alan J. (1952–) Director of tax communications at PricewaterhouseCoopers in Toronto, Canada. He coined the term *visual arithmetic* to describe the process of comparisons, estimates, and so on made by readers of graphs. An associate editor of *Information Design Journal*, he is an active promoter of good graphing technique, particularly in the business community, but also among design students and others.

deMoivre, Abraham (1667–1754) French mathematician who, as a Protestant, was unhappy in Catholic France. He moved to England (1686), where he supported himself by teaching. He is best known for his Doctrine of Chances (1718), on probability theory, and for his trigonometric formula $(\cos(\theta)) + i \sin(\theta)^n = \cos(n\theta) + i \sin(n\theta)$. This equation, by moving trigonometric functions into the complex plane, can be used to derive other trigonometric identities easily.

Donoho, David (1956–) Professor of statistics at Stanford University. Born and raised in Texas and educated at Princeton, where John Tukey was his undergraduate thesis advisor. At Princeton, Donoho became interested in robust statistics, exploratory data analysis, statistical computing, and multivariate graphics. He was one of the principal developers of MacSpin, a computer program for three-dimensional spinning of data clouds on a Macintosh computer. This was an important breakthrough, for before this such methods required a large mainframe computer. Today, Donoho works on problems in statistical inference, information theory, data compression, harmonic analysis, fast transform algorithms, and image and signal analysis. He is a member of the U.S. National Academy of Sciences and the American Academy of Arts and Sciences. He was chosen to receive a MacArthur "genius" award in 1996.

d'Oresme, Nicole (ca. 1320–1382) French mathematician, philosopher, and theologian. He was born at Allemagne in Normandy, and his academic career centered largely around the college of Navarre in the University of Paris. He left in 1361 when he was appointed first canon (1362) and then dean (1364) of Rouen. He was named chaplain to King Charles V in 1370 and elected bishop of Lisieux in November 1377 as a reward for translating a number of Aristotelian texts from Latin into French. He made a number of scientific discoveries, including suggesting the possibility of irrational proportions. He also illustrated the Merton rule of motion, a precursor to Kepler's first law, with an ingenious graph. He took a variety of stands questioning biblical dogma; for example, he cited both scientific and biblical evidence suggesting that the sun did not revolve around the Earth. For the latter, he cited Joshua 10:13, which pointed out that God made the sun stand still, so the Israelites could better slaughter their fleeing enemies. Eventually, however, he rejected heliocentrism. He is perhaps best known among mathematicians for his proof of the divergence of the harmonic series $1 + 1/2 + 1/3 + \ldots$ in his remarkable book *Quaestiones super geometriam Euclidis*, which was written around 1350. His elegant proof was independently rediscovered by Johann Bernoulli in 1697; that is the one that appears in most modern mathematics texts.

Euler, Leonhard (1707–1783) Swiss mathematician who trained under Jean Bernoulli. If "publish or perish" were literally true, Euler would still be alive. He published more than eight hundred books and papers on every aspect of pure and applied mathematics, physics, and astronomy. In 1738, when he was a professor of mathematics at St. Petersburg Academy, he lost his sight in one eye. In 1741 he moved to Berlin, but in 1766 he returned to St. Petersburg, where he soon lost his sight in the other eye. His prodigious memory allowed him to continue his work while totally blind. For the princess of Anhalt-Dessau he wrote *Lettres à une princesse d'Allemagne* (1768–1772), in which he gave a clear, nontechnical outline of the principal physical theories of the time. His *Introductio in analysin infinitorum* (1748) and later treatises on calculus and algebra remained the standard texts for more than a century.

Farquhar, Arthur Briggs (1838–1925) American businessman born in Sandy Spring, Maryland, and educated at Howells School for Boys in Alexandria, Virginia. He began his career in agriculture, running the family farm for about a year before moving to York, Pennsylvania, to train as a machinist (1856). Within two years he had become a partner, and in 1862 he began buying a greater interest in the firm. By 1889 he was president of what was now A. B. Farquhar Co. Ltd. His holdings continued to expand, and he became, at one time or another, the publisher of the *York Gazette* and president of the York Hospital, the York Municipal League, the York Oratorio Society, and the National Association of Executive Commissioners. He was also active in a wide range of other civic activities. But his only published book was *Economic and Industrial Delusions* (1891), written with his brother, Henry.

Farquhar, Henry (1851–1925) American teacher and writer born in Sandy Spring, Maryland. He was educated at Cornell University but never graduated. He began work as a teacher but then moved on to other arenas, first working for the U.S. Coast Survey, later as an editorial writer, and then as a statistician, first for the Department of Agriculture and later for the Census Bureau (1900–1921). With his brother, Arthur, he wrote *Economic and Industrial Delusions* (1891).

Ferguson, Adam (1723–1816) Scottish historian and philosopher. He became the professor of natural philosophy at Edinburgh in 1759 and of moral philosophy in 1761. He was a member of the Scottish "common sense" school of philosophy and traveled to Philadelphia as secretary of Lord North's commission to negotiate with the American colonists in 1778–1779. He wrote an *Essay on the History of Civil Society* (1767), *The History of the Progress and Termination of the Roman Republic* (1783), and *Principles of Moral and Political Science* (1792).

Fisher, Sir Ronald Aylmer (1890–1962) One of the most important developers of the connected set of mathematical techniques that have evolved to become twentieth-century statistics. A glimpse of his contributions is seen just from the number of eponymous topics in statistics: Fisher Information, Fisher Distribution, Fisher Inequality, Fisher's Exact Test, Fisher's k-Statistics, Fisher's Problem of the Nile, Fisher's z-Transformation, Fisher-Yates Tests. He is also credited with making an enormous number of other contributions that form the backbone of modern science (e.g., analysis of variance).

Franklin, Benjamin (1706–1790) Statesman, author, and scientist, who, although he was born in Boston, is most closely associated with Philadelphia, where he set up a printing house and, in 1729, bought the *Pennsylvania Gazette*. In 1737 he became postmaster of Philadelphia and in 1754 deputy postmaster general for the colonies. He was subsequently sent on various diplomatic missions to England and negotiated Britain's recognition of U.S. independence in 1783. Franklin played a major role in the framing of the Declaration of Independence (1776) and at the Federal Constitutional Convention (1787), after which he retired from public life. Until 1785, he was the U.S. minister in Paris, where he socialized with Jacques Barbeau-Dubourg and learned of Dubourg's graphical biographies. Throughout his life he was an active scientist and inventor. In 1746 he began his research into electricity and proved that lightning was electrical. He then recommended that buildings be protected by lightning conductors. Franklin is also credited with the development of improved spectacles and a wood-burning stove that provided substantial improvements in the effectiveness of home heating.

Friedman, Milton (1912–) Statistician and economist born in New York City but most firmly connected with the Chicago School of monetary theory. After eight years at the National Bureau of Economic Research (1937–1945) he joined the faculty of the University of Chicago (1946–1983), where he developed his theories of the role of money in determining events, particularly the Great Depression. He is credited with developing a method of analysis of variance based on ranks, and he won the Nobel Prize for Economics in 1976.

Funkhouser, Howard Gray (1898–1984) American mathematician and educator born in Winchester, Virginia. He was a 1921 graduate of Washington and Lee and received his Ph.D. from Columbia. He taught mathematics at Washington and Lee from 1924 to 1930 and spent 1931 on the mathematics faculty at Columbia. In 1932 he accepted a position on the faculty at Phillips Exeter Academy, where he remained until his retirement in 1966. In addition to his teaching, he remained reasonably active in scholarship. His papers "Historical Development of the Graphical Representation of Statistical Data" (1937), "Playfair and His Charts" (1935), and "A Note on a Tenth-Century Graph" (1936) together form the jumping-off point for subsequent researchers in the history of graphics.

Galilei, Galileo (1564–1642) Astronomer and mathematician. He was born in Pisa, Italy, where he trained in medicine. He became a professor of mathematics at Padua (1592–1610), where he improved the refracting telescope and used it to study the heavens. He also formulated the law of uniformly accelerated motion toward the Earth, described the parabolic path of projectiles, and recognized that all bodies have mass. He was forced to retract his advocacy of the Copernican theory in front of the Inquisition, but was nonetheless sentenced to indefinite imprisonment. At the request of Galileo's friend and protector, the Duke of Tuscany, the pope commuted his sentence; however, he remained under house arrest in Florence. The Roman Catholic Church formally recognized the validity of his scientific work in 1993.

Galton, Samuel Tertius (1783–1844) The son of Samuel Galton, he was, according to his son Francis, "eminently statistical by nature." He succeeded his father at the bank and also served as high bailiff of Birmingham. Francis noted of his father, "He wrote a small book on currency, with tables, which testifies to his taste. He had a scientific bent, having about his house the simple gear appropriate to those days, of solar microscope, orrery, telescopes, mountain barometers without which he never travelled, and so forth. A sliding rule adapted to various uses was his constant companion." Although he was devoted to Shakespeare and read *Tom Jones* every year, his duties at the bank prevented the abundant leisure he desired for systematic study. His son concluded that it was as a result of this that he became so "earnestly desirous of giving [his son] every opportunity of being educated."

Galton, Sir Francis (1822–1911) Scientist and explorer. He was born in Birmingham, England, and interrupted his medical studies at Cambridge to travel in Africa. In addition to his work on weather, he is best known for his studies of heredity and intelligence, which led to the founding of the field he dubbed eugenics. His book *Hereditary Genius* was published in 1869. His work was influential in the development of his cousin Charles Darwin's theory of evolution. He was knighted in 1909.

Gauss, Carl Frederich (1777–1855) A mathematical prodigy born in Brunswick, Germany, who wrote the first modern book on number theory. His broad contributions to virtually all areas of mathematics led Eric Temple Bell, in *Men of Mathematics*, to name him "the Prince of Mathematics." Among his other discoveries were the law of quadratic reciprocity and the intrinsic differential geometry of surfaces. He pioneered in the development of the theories of elliptic and complex functions as well as the application of mathematics within electricity, magnetism, and gravity. For his contributions the unit of magnetic inductance is named after him. He was named a professor of mathematics at Göttingen in 1807 and remained there for the rest of his life.

Ginsburg, Ruth Bader (1933–) American jurist educated at Cornell and Harvard Law School. She was born in Brooklyn, New York, and persevered through a number of hardships, including the premature deaths of her older sister and mother. At Harvard Law School, she found the atmosphere toward women extremely hostile—Dean Erwin Griswold asked the women in the class what it felt like to occupy places that could have gone to deserving men. She taught at Rutgers (1963), where she championed a number of liberal causes, including maternity leave for schoolteachers in New Jersey. She was the first woman hired with tenure at Columbia Law School (1972), the same year she was selected as director of the American Civil Liberties Union's Women's Rights Project. She argued six cases for women's rights before the Supreme Court before being named an associate justice of that court in 1991 by President Bill Clinton.

Glazer, Nathan (1924–) American sociologist and educator. He was born in New York City and completed his Ph.D. at Columbia University. Currently, he is professor emeritus at Harvard University's Graduate School of Education. Glazer is the coeditor of *The Public Interest* and a leading authority on issues of race and social policy in the United States. His books include *The Melting Pot* (1963) and, with David Riesman and Reuel Denney, *The Lonely Crowd* (1950). He served on presidential task forces on education and urban policy and the National Academy of Science's committees on urban policy and minority issues. He is also a contributing editor of the *New Republic*.

Graunt, John (1620–1674) British statistician. Received a rudimentary British education and was apprenticed to a haberdasher, a trade he followed for most of his life. He published the first analysis of the Bills of Mortality in 1662 in the effusively titled *Natural and Political Observations Mentioned in a Following Index, and Made upon the Bills of Mortality, by John Graunt, Citizen of London, with Reference to the Government, Ayre, Diseases, and the Several Changes of the Said City*. His analysis of the data surrounding the deaths of Londoners included the preparation of life-expectancy tables that indicated the proportion of persons who could be expected to survive to various ages. Such tables form the actuarial basis of modern life insurance. Even though Graunt was a shopkeeper, he was immediately offered membership in the Royal Society. His membership was recommended by Charles II, who added "that if they found any more such tradesmen, they should be sure to admit them all without any more adoe."

Hilbert, David (1862–1943) Mathematician who was born in Königsberg, studied mathematics there, and became a professor there (1893). Two years later, he moved to Göttingen, where he made important contributions to the foundations of geometry, number theory, the theory of invariants, and algebraic geometry. At the International Congress of Mathematics in 1900, he gave a talk in which he listed what he believed were the twenty-three most important unsolved problems in mathematics. These problems, called thereafter the Hilbert Problems, formed the basis of much mathematical research in the subsequent century.

Hipparchus (second century B.C.) Greek astronomer, born in Nicaea, Rhodes, who carried out observations that allowed him to discover the precession of the equinoxes as well as the eccentricities of the sun's path. He estimated (within seven minutes!) the length of the solar year as well as the distances to the sun and the moon from the Earth. He also cataloged 1,080 stars. He was able to fix the position of places by their longitude and latitude, and invented trigonometry.

Holland, Paul (1940–) American statistician born in Oklahoma but raised in Michigan. He was trained at Stanford and taught at Harvard and Berkeley. Most of his career was spent at the Educational Testing Service (1975–1993; 2000–), and during his tenure there he made fundamental contributions to the theory of testing, discrete multivariate analysis, differential item functioning, and item response theory. He also learned to play the banjo from Jerry Garcia.

Hotelling, Harold (1895–1973) Statistician and economist who was born in Fulda, Minnesota, but moved to Seattle as a child. His undergraduate education at the University of Washington was interrupted by his service in World War I. He studied journalism but found he wrote too slowly to pursue that as a career, and so he returned to Washington for a master's degree in mathematics. After being rejected by Columbia for further graduate study in economics, he attended Princeton, where he hoped to learn more about both statistics and mathematical economics. Unfortunately, as he was to write later, he was unable to find anyone on the faculty "who know anything about either subject." He completed a Ph.D. in mathematics in 1924. After teaching at Stanford for seven years, he accepted a position in economics at Columbia, where he remained for fifteen years. He left Columbia to found the Department of Statistics at the University of North Carolina, where he remained until his retirement in 1966. His work spans both statistics and economics. He was a major contributor to the development of multivariate analysis. Within economics he wrote seminal papers on spatial competition, the economics of exhaustible resources, and marginal cost pricing. He showed that the rate of increase in the unit value of the remaining stock of a depletable resource will equal the rate of interest under competition, which has since been called Hotelling's Rule.

Howard, Luke (1772–1864) British businessman and unlettered meteorologist. His career in meteorology began when he was eleven and noticed the spectacular overhead show caused by a lingering cloud of dust that originated in volcanic activity in Japan and Iceland. He ended up devising a taxonomy of clouds that he named. His names combined a component for type (*cumulus*, *stratus*, *cirrus*, and *nimbus*) and for height (*alto* and *cirro*). In 1800 he presented his names before the Linnaean Society of London, and was named a Fellow of the Royal Society in 1821. He was also the first to notice that cities have an effect on the climate. His observation that cities were often warmer than the surrounding countryside is now known as the urban heat island effect.

Hume, David (1711–1776) Philosopher and historian born in Edinburgh, where he studied law. In 1734, he went to La Fléche in Anjou, where he wrote *A Treatise of Human Nature* (1740), which extended and consolidated the empiricism of John Locke and Bishop Berkeley. He returned to Edinburgh, although as a tutor, secretary, and keeper of the Advocates' Library, not as a professor, as his applications for such professorial positions were turned down because of his atheism. He published his popular *Political Discourses* in 1752, and a six-volume *History of England* over the eight-year period 1754–1862. Hume's views inspired Kant's arguments on the inadequacy of empiricism.

Huygens, Christiaan (1629–1695) Physicist and astronomer born in The Hague who studied in Leiden and Breda. He discovered the rings of Saturn as well as its fourth moon (1655). A graphic depiction of motion led him to understand that the period of a pendulum is related only to its length and not to its amplitude. This led him to build the first pendulum clock (1657). He was a major contributor to the wave theory of light and discovered polarization. He lived principally in Paris and was a member of the Royal Academy of Sciences (1666–1681). He returned to The Hague when, as a Protestant, he decided that prudence dictated.

Jefferson, Thomas (1743–1826) Farmer, statesman, and third president of the United States (1801–1809). He was born in Shadwell, Virginia, and studied at the College of William and Mary. He became a lawyer (1767) and took a prominent role in the first Continental Congress (1774), where he drafted the Declaration of Independence. He served as governor of Virginia (1779–1781), from 1784 to 1789 was ambassador to France (where he, too, met Jacques Barbeau-Dubourg), was secretary of state (1789), and was vice-president under John Adams (1797–1801). During his presidential administration, the country went to war with Tripoli, completed the Louisiana Purchase, and outlawed the slave trade. He retired from public life in 1809 but remained active as an architect, scientist, educator, and senior advisor on matters of state. He died on July 4, 1826, fifty years to the day after the publication of the Declaration of Independence.

Johnson, Samuel (1709–1784) Lexicographer, critic, and poet, He was born in Lichfield, England, and studied at Lichfield and Oxford but never completed his degree. He was known broadly as "Dr. Johnson" for reasons unrelated to his formal education. He went to London in 1737 and worked as a journalist. In 1747 he began the eight-year-long task of compiling his *Dictionary of the English Language*. This task, apparently, did not fully occupy him, for in 1750 he inaugurated the moralistic periodical *The Rambler*. In 1762 he was given a crown pension, which allowed him the time to become a critic and society personality. He was a founding member of the Literary Club in 1764, after which he was quite active producing his edition of Shakespeare in 1765, as well as a multitude of pamphlets (1772). In 1773 he toured Scotland with James Boswell and later wrote *Lives of the Poets* (1779–1781). Most of what we know about Dr. Johnson we learned from Boswell's *Life of Samuel Johnson*.

Kames, Lord (Henry Homes) (1696–1782) Lord Kames was a commoner who took his title after being seated as a judge on the Court of Session and styled himself Kames after his family estate in Kames, Berwickshire. He was a prolific essayist and confidant of David Hume and James Boswell. In 1737, during his time as a senior examiner for the Faculty of Advocates, he was named curator of the Advocates' Library. Within five years, he turned it into a major repository for books on law, philosophy, history, and geography. This library was a crucial aid for Hume in the writing of his *History of England*, and for Adam Ferguson in the writing of his *Essay on the History of Civil Society*.

Kelley, Truman Lee (1884–1961) Educator born in Michigan, who studied mathematics at the University of Illinois. He was an instructor in mathematics at Georgia Institute of Technology and at the Hish School in Fresno, California, before returning to graduate school to study psychology. He got a master's degree at the University of Illinois in 1909 and a Ph.D. at Columbia in 1914, where he was introduced to psychometrics. He taught at the University of Texas (1914–1917) and at Columbia (1917–1920) before moving to Stanford as an assistant professor of education. There he interacted with Louis Terman and helped develop the Stanford Achievement Test Battery, writing *Statistical Methods* in 1923 and *Interpretation of Educational Measures* in 1927, and worked with Hotelling on factor analysis. He subsequently taught at Harvard (1931–1950) in the Graduate School of Education.

Kepler, Johannes (1571–1630) Astronomer born in Weil-der-Stadt, Germany. He studied at Tübingen and was appointed as a professor of mathematics in Graz in 1593. In 1596 he began to correspond with Tycho Brahe, who was then in Prague, and began a collaboration that yielded the eponymous laws governing planetary motion. Kepler's first and second laws appeared in *Astronomia nova* (1609) and formed the basis of Newton's laws of motion. Kepler's third law appeared in *Harmonice mundi* (1619). He followed Brahe as court astronomer to Emperor Rudolf II, and in 1628 he became astrologer to Albrecht Wallenstein.

King, Willford I. (1880–1962) American economist born in Cascade, Iowa. He was educated in the Midwest, first at the University of Nebraska and then, for his Ph.D., at the University of Wisconsin (1910), where he stayed on as an assistant professor until 1917. Dr. King worked as a statistician for the U.S. Public Health Service until 1920 and as an economist at the National Bureau of Economic Research until 1927, when he joined the economics faculty of New York University, where he remained until his retirement in 1945. King was best known for his advocacy of a sliding scale of wages based on productivity. He was a vehement opponent of the New Deal, despite the concern he had expressed earlier about the negative consequences of three-fifths of the property in the United States being owned by a mere 2 percent of the people. Early in his career he advocated "little socialist states within the capitalist states," in which workers could be paid in scrip. He overcame these socialist tendencies later in life and described the pay raise demands of the Congress of Industrial Organizations as a threat to the manufacturing industry. In testimony before the House Ways and Means Committee, he called for a reduction of taxes in the upper brackets and the abolition of all taxes on corporate income and on income from invested capital.

Lagrange, Joseph Louis, Comte de l'Empire (originally Giuseppe Luigi Lagrangia) (1736–1813) Mathematician and astronomer, born in Turin, Italy. At the age of thirty he was named director of the Berlin Academy, where he published extensively on number theory, mechanics, algebraic equations, and celestial mechanics. His principal work was *Mécanique analytique* (1788), after which he was appointed professor of mathematics at the École Polytechnique, where he headed the committee that reformed the metric system (1795). He was named a senator and made a count by Napoleon. The Lagrangian point in astronomy, the Lagrangian function in mechanics, and the Lagrange multiplier in mathematics are all part of his intellectual legacy.

Laplace, Pierre Simon, Marquis de (1749–1827) Mathematician and astronomer born in Beaumont-en-Auge, France. After completing his studies at Caen, he became a professor of mathematics at the École Militaire in Paris. Although his research was very wide-reaching, his primary focus was on the application of mathematics in celestial mechanics. His five-volume *Mécanique céleste* (1799–1825) remains a landmark in applied mathematics. In his examination of the gravitational attraction of planets he posited the fundamental differential equation in physics that now bears his name. He entered the senate in 1799 and was made a marquis in 1815. It is said that Napoleon met with him and said that he, Napoleon, was disappointed in reading Laplace's great opus to find "the name of God not mentioned even once." Laplace is said to have replied, "I did not need that hypothesis." (There is yet more to this story. Apparently, when Napolean related this interaction to Lagrange, who was thirteen years older than Laplace, Lagrange is said to have replied sadly, "Too bad. It is a wonderful hypothesis and accounts for many things.")

Legendre, Adrien-Marie (1752–1833) French mathematician. After his studies at the Collége Mazarinche, he became a professor of mathematics at the École Militaire (1775–1780), a member of the Académie des Sciences (1783), and a professor at the École Normale (1795). He made major contributions to number theory and elliptical functions.

Lister, Martin (ca. 1600) Apparently the first user of what he called a "compendious way" of observing the barometer. On March 10, 1683, Lister presented to the Oxford Philosophical Society a graph on which wind directions, barometric pressures, and temperatures appear side by side. Lister's graphs were done tediously by hand.

Marey, Etienne Jules (1830–1903) Physiologist and ardent believer in the power of the visual system to discover truths and understand nature. He was born in Beaune, France, and at age thirty-five became a professor at the Collége de France. His studies of animal movement (1887–1900) were pathbreaking uses of scientific cinematography. His innovations in camera design resulted in reducing exposure time to about 1/25,000 of a second to photograph the flight of insects. He also designed the finest train schedule ever produced (see figure I.1).

Meikle, Andrew (1719–1811) Millwright and inventor, born in East Lothian, Scotland. He inherited his father's mill, and to improve production invented the fantail (1750), a machine for dressing grain (1768), and the spring sail (1772). His most important invention was a drum threshing machine (patented in 1788), which could be driven by wind, water, horse, or (some years later) steam power.

Michelson, Stephan (1938–) President of Longbranch Research Associates. He was educated at Stanford (1968) in economics and subsequently taught at Reed College and Harvard. In 1979 he cofounded Econometric Research, Inc. (Washington, DC), which grew to become Longbranch Research Associates. He is a specialist in statistical analysis in litigation and introduced into the legal process the concepts of using multiple pools and survival analysis for selection issues, as well as process analysis for jury composition issues.

Minard, Charles Joseph (1781–1870) He was first a civil engineer and then an instructor at the École Nationale des Ponts et Chaussées. He later was an inspector general of the Council des ponts et chaussées, but his lasting fame derived from his development of thematic maps in which he overlaid statistical information on a geographic background. The originality, quality, and quantity of this work led some to call him "the Playfair of France" (Funkhouser 1937). His intellectual leadership led to the publication of a series of graphic reports by the Bureau de la statistique graphique of France's Ministry of Public Works. The series (*L'Album de statistique graphique*) continued annually from 1879 until 1899 and contained important data on commerce that the ministry was responsible for gathering. In 1846, he developed a graphical metaphor of a river, whose width was proportional to the amount of materials being depicted (e.g., freight, immigrants), flowing from one geographic region to another. He used this almost exclusively to portray the transport of goods by water or land. This metaphor was employed to perfection in his 1869 graphical depiction of the flow of Napoleon's troops in their doomed march into Moscow and back in the winter of 1812. The rushing river of 422,000 men that crossed into Russia, when compared with the returning trickle of 10,000, "seemed to defy the pen of the historian by its brutal eloquence." The graph carries seven variables clearly. This now-famous display has been called (Tufte 1983) "the best graph ever produced."

Molyneux, William (1656–1698) Irish philosopher born in Dublin, the eldest surviving son of Samuel Molyneux (1616–1693), a wealthy property owner. He was originally trained at Trinity College, Dublin, before moving to London to study law at the Middle Temple. He had little interest in law and focused instead on philosophy and mathematics, returning to Dublin in 1678. Over the course of his life he worked in an amazing number of different areas. His *Dioptrica nova* (1692) was for a long time considered the standard work in optics. This grew out of his work in astronomy, which was marked, in 1686, by the publication of *Sciothericum Telescopicum: or, A New Contrivance of Adapting a Telescope to a Horizontal Dial.* He is best known for his 1698 work *The Case of Ireland's Being Bound by Acts of Parliament in England Stated.* Among his many collaborators were Edmund Halley and John Locke. He spent two terms as Trinity College's representative in Parliament.

Mosteller, Fred (1916–) American statistician born in Clarksburg, West Virginia. He was John Tukey's first (and perhaps most eminent) graduate student. After completing his Ph.D. (1946) he emigrated to Harvard, where he has remained ever since, chairing Harvard's departments of social relations (1953–1954), statistics (1957–1969, 1973, 1975–1977), biostatistics (1977–1981), and health policy and management (1981–1987). He has been so prolific and his work so broad, ranging over so many different areas, that it is impossible to characterize it accurately in just a few words. There are essentially no topics in social science or education to which he has not made a contribution. His publication list includes fifty-eight books and more than 350 articles and book chapters. Although there is no formal position called "statistician general of the United States," Frederick Mosteller has held it informally since about 1960.

Musgrave, William (1655–1721) Physician and antiquary born in Nettlecombe, Somerset, England. He was educated at Oxford and, for his distinction in natural philosophy and physics, was elected to the Royal Society in 1684. The following year he served as secretary of the Society and edited the *Philosophical Transactions* from volume 167 to 178. His best-known treatises are *De arthritide symptomatica* in 1703, followed by *De arthritide anomala* in 1707.

Nightingale, Florence (1820–1910) Hospital reformer and graphic designer. She was born in Florence, Italy, but raised in England. She trained as a nurse in Kaiserswerth and Paris, and in the Crimean War, after the Battle of Alma (1854), she organized a party of thirty-eight nurses to be the nursing department at Scutari. There she found grossly inadequate sanitation but soon improved matters, which was a good thing, for she soon had more than ten thousand wounded men under her care. In 1856 she returned to England, where she formed a training program for nurses at St. Thomas Hospital. She spent a number of years campaigning for army sanitary reform, general improvements in nursing, and instituting modern public health policies in India.

O'Connor, Sandra Day (1930–) Jurist, the first female justice on the U.S. Supreme Court. She was born in El Paso, Texas, but studied law at Stanford and was admitted to the bar in California. She held several elected and appointed positions in Arizona before being named as an associate justice to the Supreme Court in 1981.

Playfair, John (1748–1819) Minister, geologist, and mathematician. Born in Dundee, Scotland, he studied at St. Andrews and became a professor of mathematics (1785) and natural philosophy (1805) at Edinburgh University. In addition to writing a textbook on geometry, he also investigated glaciation and the formation of river valleys. He was the older brother of William Playfair.

Playfair, William (1759–1823) Scottish iconoclast, the father of modern graphical methods. See chapters 1–4.

Pliny (the Elder), in full Gaius Plinius Caecilius Secundus (23–79) Roman scholar, born in Como, Gaul. He studied at Rome, served in the army in Germany, and settled in Como. Nero appointed him procurator in Spain. He wrote a thirty-seven-volume encyclopedia, *Historia naturalis* (77), which is his only work to survive. When he was fifty-six years old, he was in command of the Roman fleet when the great eruption of Vesuvius was at its peak. He landed at Castellamare to observe more closely, and was killed by volcanic fumes.

Plot, Robert (1640–1696) Alchemist and scientific dilettante born in Kent, England. He was born into wealth and educated in natural history at Oxford. He planned to write a natural history of England and Wales and began with the *Natural History of Oxfordshire* in 1677, which led to his election to the Royal

Society. He subsequently completed the *Natural History of Staffordshire*, but stumbled when he worked on those of Kent and Middlesex. His interests in curiosities and antiquities apparently derailed him. He served as a professor of chemistry at Oxford (1683–1690), but his principal interest was in pursuit of a universal solvent.

Priestley, Joseph (1733–1804) Chemist and clergyman, born in Fieldhead, West Yorkshire, England. He became a Presbyterian minister in 1755 and moved to Leeds in 1767, whereupon he took up the study of chemistry. He is best known for his research on the chemistry of gases and for his discovery of oxygen. Along with books on education and politics he also wrote an English grammar. His support of the French Revolution was controversial, and so in 1794, in fear for his life, he emigrated to America, where he was well received.

Santayana, George, originally Jorge Augustin Nicolas Ruiz de Santayana (1863–1952) Philosopher, poet, and novelist. He was born in Madrid but moved to Boston when he was nine. He was educated at Harvard, where he was a professor of philosophy (1907–1912). His writing career began as a poet with *Sonnets and Other Verses* (1894), but his later works (*The Life of Reason*, five volumes, 1905–1906; *Realms of Being*, four volumes, 1927–1940) were in philosophy, although he did venture into fiction as well (*The Last Puritan*, 1935). He moved to Oxford during World War I and then settled in Rome.

Scalia, Antonin (1936–) American jurist born in Trenton, New Jersey. The only child of an immigrant Italian family, he married in 1960 and subsequently fathered nine children. He was educated at Georgetown and Harvard Law (1960), after which he briefly worked in commercial law before joining the faculty at the University of Virginia. His subsequent career alternated between stints of government work and the academy. During an appointment at the law school of the University of Chicago, he bought an old fraternity house to contain his family. He was appointed to the U.S. Court of Appeals for Washington, DC, by President Reagan (1982), and when Warren Burger retired in 1986, Reagan nominated Scalia to the Supreme Court. On the court his reputation is that of the "intellectual conservative." He has been a staunch opponent of the reproductive rights asserted in *Roe v. Wade* (1973) and has consistently rejected any right to abortion under the Constitution.

Schmid, Calvin (1902–1994) Sociologist and demographer. A native of Dayton, Ohio, he was trained at the University of Washington, where he spent virtually his entire professional career. His 1983 book *Statistical Graphics* kept alive a set of procedures for displaying data that was dead before the book was conceived.

Secrist, Horace (1881–1958) American economist. He was born in Farmington, Utah, but grew up in Provo. He was educated at the University of

Wisconsin, completing a Ph.D. in 1911. He taught economics at Wisconsin until 1918, when he became director of Northwestern University's Bureau of Business Research, a position he held until 1933, although he had brief sojourns in other places during his tenure. Among the ten books he published were *Statistics in Business* (1920), *The Widening Retail Market* (1926), *Banking Ratios* (1930), and the infamous *Triumph of Mediocrity in Business* (1933).

Seneca, Lucius Annaeus (the Elder) (ca. 55 B.C.–40 A.D.) Roman rhetorician born in Cordoba, Spain. He wrote a history of Rome (now lost) as well as several works on oratory. Parts of his *Colores controversiae* and *Suasoriae* have survived.

Sharpe, William F. (1934–) American economist trained at UCLA, who made his reputation in the development of the capital asset pricing model, the Sharpe ratio for investment performance analysis, the binomial method for the valuation of options, and the gradient method for asset allocation optimization. He is the author of six books, including *Portfolio Theory* and *Capital Markets* (1970) and *Asset Allocation Tools* (1987). He is currently the STANCO 25 Professor of Finance Emeritus at Stanford University, as well as chairman of Financial Engines, Inc. In 1990 he received the Nobel Prize in Economics.

Smith, Adam (1723–1790) Economist and philosopher. He was born in Kirkcaldy, Fife, Scotland, and studied at Glasgow and Oxford. He became a professor of logic at Glasgow in 1751 and of moral philosophy in 1752. In 1776 he moved to London and published *An Inquiry into the Nature and Causes of the Wealth of Nations*, which was the first major work on political economics. He moved to Edinburgh in 1778 when he was appointed commissioner of customs there.

Snow, John (1813–1858) Anesthesiologist and epidemiologist born in York. He practiced medicine in London from 1836, and during cholera outbreaks of 1848 and 1854 carried out epidemiological investigations. These culminated in the construction of his famous map from which he deduced that the Broad Street pump suffered from seepage of sewage. He was also a pioneer anesthesiologist who experimented with ether and chloroform and devised an apparatus to administer them. As physician to Queen Victoria in 1853, he administered chloroform to her during the birth of Prince Leopold.

Souter, David (1939–) American jurist born in Melrose, Massachusetts, and educated at Harvard. He was a Rhodes scholar, spending two years at Magdalen College, Oxford. He practiced law in New Hampshire, but his political activism led to his eventually being appointed attorney general of New Hampshire (1976). He was subsequently promoted to the New Hampshire Supreme Court (1983) and in 1990 to the U.S. Court of Appeals. When Justice William J. Brennan retired five months later, President Bush nominated him to be an associate justice of the U.S. Supreme Court.

Spence, Ian (1944–) Scottish-Canadian psychologist born in Glasgow, who received his undergraduate education at the University of Glasgow. He emigrated to Canada in 1966 and completed his Ph.D. at the University of Toronto (1970), where he is currently director of the Government Research Infrastructure Programs Office and a professor in the Department of Psychology. He has published more than sixty papers and monographs in engineering psychology, graphical perception, psychophysics, psychometric methods, and statistical graphics. His Scottish heritage and work on graphs inspired his interest in William Playfair, on whose biography he is currently working.

Stamp, Sir Josiah Charles (1880–1941) London-born British economist who served on the Dawes Committee on German reparations after World War I and was killed in an air raid during World War II. He was also chairman of the London, Midland & Scottish Railway.

Stevens, John Paul (1920–) American jurist born and educated in Chicago. After service in the navy during World War II, he followed in his father's footsteps and attended Northwestern University's law school. He then joined a prominent Chicago law firm and specialized in antitrust law. In 1970, Richard Nixon appointed him to the Seventh Circuit U.S. Court of Appeals. In 1975, Attorney General Edward Levi, former president of the University of Chicago, recommended him to fill the vacancy on the Supreme Court left by the retirement of Justice William O. Douglas. President Gerald Ford nominated him and he was confirmed by the Senate without controversy.

Stewart, Potter (1915–1985) American jurist born in Ohio and educated at Yale. After law school and service in the navy during World War II, he followed his father's lead in entering Cincinnati politics. He was elected to the city council in 1950 and as vice-mayor in 1952 (his father had served as mayor of Cincinnati many years earlier). He was nominated by President Eisenhower to become an associate justice of the U.S. Supreme Court in 1958, where he served for twenty-three years before retiring to Hanover, New Hampshire. He was a centrist on the court and is best remembered for his concurrence in the pornography case *Jacobellis v. Ohio*. He was unable to provide a definition of pornography but, he declared, "I know it when I see it."

Stigler, Stephen (1941–) Statistician and historian. He was trained at Stanford and is currently the Ernest DeWitt Burton Distinguished Service Professor of Statistics at the University of Chicago. Among his written works he is best known for his *History of Statistics: The Measurement of Uncertainty before 1900* (1986) and *Statistics on the Table* (1999). He is the son of the Nobel Prize–winning economist George Stigler.

Stuart, Gilbert (1755–1828) Painter born in Rhode Island. At the age of twenty, he went to London to study portraiture under Benjamin West. He became a popular portrait painter and in 1792 returned to America, where he painted almost a thousand portraits, including those of Washington, Jefferson, Madison, and John Adams.

Thomas, Clarence (1948–) Jurist born in Savannah, Georgia, who studied at Yale and was named by President Bush to be the second black American to sit on the Supreme Court. His confirmation hearings were dominated by charges of sexual misconduct brought by Anita Hill, a former aide, but he was confirmed nonetheless (1992). His performance on the high court has been consistently right-wing and unexceptional.

Thoreau, Henry David (1817–1862) Essayist and poet born in Concord, Massachusetts, and educated at Harvard. He worked as a teacher and in his late twenties began taking long walks and studying nature. He sometimes lived at the family home of Ralph Waldo Emerson (1803–1882), but from 1845 until 1847 he lived in a shack he built at Walden Pond (near Concord). While in residence at Walden Pond, he wrote his most famous work, *Walden, or Life in the Woods*, which was published in 1854. He wrote and lectured broadly and his advocacy of individualism, fitting as well as it did into the American ethos, was highly influential.

Thucydides (ca. 460–400 B.C.) Historian of the Peloponnesian War. He was an Athenian who provided an accurate narrative of the war. He was critical of the democratic system's performance during the war years and was exiled in 424 for twenty years by the democracy, which accused him of military incompetence in the Aegean.

Tufte, Edward R. (1942–) American political scientist and graphics expert. He was born in California and trained at Yale and Stanford. The seven books he has published include *The Visual Display of Quantitative Information* (1983), *Envisioning Information* (1990), and *Visual Explanations* (1996). These books have received unprecedented attention, garnering among them more than forty awards for content and design. They have sold, in the aggregate, more than a half-million copies and in the process have made Tufte both rich and famous.

Tukey, John Wilder (1915–2000) American polymath who made broad contributions to science and was the principal developer and proselytizer of modern exploratory data analysis. He coined the words *bit*, *software*, *ANOVA*, and *stem-and-leaf diagram*. See chapter 19.

Von Humboldt, Alexander (1769–1859) Naturalist and geographer born in Berlin and educated in Frankfurt, Berlin, Göttingen, and Freiberg. He spent 1799 through 1804 exploring South America with Aimé Bonpland (1773–1858). When he was fifty-eight, he spent three years traveling throughout central Asia. The Pacific current of the coast of South America is named for him. His principal book, *Kosmos*, tries to provide a comprehensive characterization of the universe.

Wallace, David L. (1930–) American statistician trained at Princeton. He was Tukey's eighth Ph.D. student, obtaining his degree in 1954. He spent most of his career at the University of Chicago. He is best known for his work on principled methods of formal statistical inference, especially Bayesian, fiducial, and likelihood-based methods, and their relation to informal methods of exploratory data analysis. His 1964 book with Frederick Mosteller, *Inference and Disputed Authorship: The Federalist Papers*, was a statistical tour de force that showed how discriminant analysis and other statistical methods can be usefully employed to settle issues of disputed authorship.

Watt, James (1736–1819) Inventor born in Greenock, Scotland. In 1754 he went to Glasgow to apprentice in the trade of instrument maker, and then stayed and set up a business in the city. As part of his trade he did surveys for canals, and began to study the use of steam as an energy source. In 1763 he repaired a model Newcomen steam engine and found he could improve its efficiency through the use of a separate steam condenser. He joined together with Mathew Boulton and began manufacturing an improved steam engine in Birmingham in 1774. He subsequently made numerous other inventions, including the double-acting engine, parallel motion linkage, the centrifugal governor for automatic speed control, and the pressure gauge. He is credited with coining the term *horse-power*. The standard unit of electrical power is named for him.

Westbrooke, Ian (1954–) New Zealand statistician trained at UCLA and the University of Wellington in New Zealand. He currently works at Statistics New Zealand on ethnic and age patterns of marriage.

Wolsey, Thomas D. (ca. 1473–1530) Cardinal, archbishop of York, Lord Chancellor, and Henry VIII's chief minister from 1515 until 1529. He was born in Ipswich, Suffolk, and studied at Oxford. He was ordained in 1498 and appointed chaplain to Henry VII in 1507. Under Henry VIII he pursued legal and administrative reforms and was in charge of the day-to-day running of the government. His high taxation to satisfy the needs of royal foreign policy caused considerable resentment, and when he was unsuccessful in persuading the pope to annul Henry's marriage to Catherine of Aragon, he was impeached and some of his property forfeited. He was allowed to retire to his archbishopric of York. He had never been to York in the fifteen years since he had been made archbishop, and indeed, he never quite made it. On the trip there, which he took at a very slow pace, he was involved in an indiscreet correspondence with Rome. He was arrested on a charge of high treason. He died in Leicester, while en route to London.

Wren, Sir Christopher (1632–1723) Architect and astronomer born in East Knoyle, England. He was trained at Oxford and became a professor of astronomy there (1661). After the great fire of London in 1666 he prepared a design for rebuilding the whole city. It was never adopted, but his design for St. Paul's Cathedral was (he is buried there). He also designed many other public buildings, such as the Greenwich Observatory and the Royal Exchange. He was one of the founders of the Royal Society and was knighted in 1673. He was also a member of Parliament (1685).

Zorba, Alexis (1883–1957) The title character in Nikos Kazantzakis's novel *Zorba the Greek* (1946), whose joie de vivre was made famous by Anthony Quinn's Oscar-winning portrayal in the 1964 movie of the same name.

Notes

Part I: William Playfair and the Origins of Graphical Display

1. Marey 1878, p. iii.
2. Ibid., p. vi.
3. Daston and Galison 1992, p. 81.
4. Ibid.

Chapter 1: Why Playfair?

1. Beniger and Robyn 1978; Costigan-Eaves 1984; Funkhouser 1937; Funkhouser and Walker 1935; Marey 1885; and Tilling 1975.
2. Smith 1925.
3. Apel 1944.
4. This was discovered by Patricia Costigan-Eaves and Michael Macdonald-Ross and described in their as yet unpublished monograph on the history of graphics.
5. Beniger and Robyn 1978; and Funkhouser 1937.
6. Biderman 1978.
7. Costigan-Eaves and Macdonald-Ross (in progress).
8. Priestley 1769.

Chapter 2: Who Was Playfair?

1. Playfair's unpublished "Reflections on His Life," transcribed, edited, and annotated by Ian Spence.

Chapter 3: William Playfair: A Daring Worthless Fellow

1. Herschel 1833.
2. Fitzgerald 1904; and De la Torre 1952.
3. Funkhouser 1937.

Chapter 4: Scaling the Heights (and Widths)

1. Road shaped as a K, by Peter Porges. Reproduced by permission of The Cartoon Bank. The cartoon on p. 31, "Miss Harper . . . let me know the minute anything happens," is by Thomas W. Cheney.

2. Brinton 1914, p. 352.
3. Schmid 1983, p. 28.
4. Cleveland 1994, p. 81.

Chapter 7: The Graphical Inventions of Dubourg and Ferguson

1. Bertin 1973 (English translation is Bertin 1983).

Chapter 8: Winds across Europe

1. Galton 1863a.
2. Ibid., p. 386.

Chapter 10: Two Mind-Bending Statistical Paradoxes

1. Westbrooke 1998.
2. Jeon, Chung, and Bae 1987.
3. Mills 1924, p. 394.
4. Stigler 1997, p. 112.
5. Ibid.
6. Secrist 1933.
7. Hotelling 1933, p. 164.
8. King 1934.
9. Sharpe 1985, p. 430.
10. Friedman 1992.
11. Kelley 1927.
12. Linn 1982; and Reilly 1973.

Chapter 11: Order in the Court

1. From Jacob Bronowski's TV series, *The Ascent of Man.*

Chapter 16: There They Go Again!

1. Wainer 1980a and 1997b.

Chapter 19: John Wilder Tukey

1. This paragraph is a close paraphrasing of Brillinger 2002, p. 193.
2. Thompson 1961.
3. Cleveland 1994, pp. 97–98.
4. Burt 1961.

5. Conway 1959.
6. Dorfman 1978.
7. Hearnshaw 1979, p. 245: "Conway [was one of the] members of a large family of characters invented to save [Burt's] face and boost his ego."

Chapter 20: Graphical Tools for the Twenty-first Century: I. Spinning and Slicing

1. Wainer and Thissen 1977.
2. Tukey 1977.
3. For a fuller story about Galileo and his time, see the charming book by Dava Sobel (1999).

Chapter 21: Graphical Tools for the Twenty-first Century: II. Nearness and Smoothing Engines

1. For fuller stories of John Snow's investigation, see Gilbert 1958 and Tufte 1996.
2. Randolph 1934.

Chapter 22: Epilogue

1. Wainer, Palmer, and Bradlow 1998.
2. Newrick, Affie, and Corrall 1990.
3. Mangel and Samaniego 1984; Wald 1980.
4. Dynarski 1987; Edwards and Beckworth 1990; Page and Feifs 1985; Powell and Steelman 1984; Steelman and Powell 1985; Taube and Linden 1989.
5. For such details the interested reader is referred to Little and Rubin 1987; Rosenbaum 1995; and Wainer 1986 for a beginning.

References

Anderson, E. 1928. The problem of species in the northern blue flags, *Iris versicolor L.* and *Iris virginica L. Annals of the Botantical Garden* 15: 241–332.

Angoff, C., and H. L. Mencken. 1931. The worst American state. *American Mercury* 31: 1–16, 175–188, 355–371.

Apel, W. 1944. *The notation of polyphonic music.* Cambridge, MA: Mediaeval Academy of America.

Arbuthnot, J. 1710. An argument for divine providence taken from the constant regularity in the births of both sexes. *Philosophical Transactions of the Royal Society* (London) 27: 186–190.

Asimov, D. 1982. The grand tour. A talk given at the Advanced Workshop on Graphics, Stanford University, Stanford, California, July.

Baker, S. G., and B. S. Kramer. 2001. Good for women, good for men, bad for people: Simpson's paradox and the importance of sex-specific analysis in observational studies. *Journal of Women's Health and Gender-Based Medicine* 10: 867–872.

Beniger, J. R., and D. L. Robyn. 1978. Quantitative graphics in statistics: A brief history. *American Statistician* 32: 1–10.

Berry, S. M. 2002. One modern man or 15 Tarzans? *Chance* 15 (2): 49–53.

Bertin, J. 1973. *Semiologie graphique.* 2d edition. The Hague: Mouton-Gautier. English translation: see Bertin 1983.

———. 1981 *Graphics and graphic information processing.* Trans. William Berg and Paul Scott; technical ed. Howard Wainer. New York: Walter de Gruyter. English translation of the 1977 edition of *La graphique et le traitement graphique de l'information.*

———. 1983. *Semiology of graphics.* Trans. William Berg; technical ed. Howard Wainer. Madison: University of Wisconsin Press. English translation of Bertin 1973.

Biderman, A. D. 1978. Intellectual impediments to the development and diffusion of statistical graphics, 1637–1980. Paper presented at first General Conference on Social Graphics, Leesburg, VA.

———. 1990. The Playfair enigma: The development of the schematic representation of statistics. *Information Design Journal* 6 (1): 3–25.

Bogert, B. P., M. J. R. Healy, and J. W. Tukey. 1963. The quefrency alanysis of time series for echoes: cepstrum, pseudo-autocovariance: Cross-cepstrum and saphe-cracking. In *Proceedings of the symposium on time series analysis,* ed. Murray Rosenblatt, pp. 209–243. New York: John Wiley.

Bowen, W. G., and D. Bok. 1998. *The shape of the river.* Princeton, NJ: Princeton University Press.

Brakenridge, W. 1755. A letter from the Reverend William Brakenridge, D.D. and F.R.S., to George Lewis Scott, Esq., F.R.S., concerning the London *Bills of mortality. Philosophical Transactions of the Royal Society* (London) 48: 788–800.

Brillinger, D. R. 2002. John Wilder Tukey (1915–2000). *Notices of the AMS* 49 (2): 193–201.

Brinton, W. C. 1914. *Graphic methods for presenting facts.* New York: Engineering Magazine Company.

Browder, F., ed. 1976. *Mathematical developments arising from Hilbert problems.* Symposia in Pure Mathematics, vol. 28. Providence: American Mathematical Society.

Burt, C. L. 1961. Intelligence and social mobility. *British Journal of Statistical Psychology* 14: 3–23.

Clagett, M. 1968. *Nicole Oresme and the medieval geometry of qualities and motions.* Madison: University of Wisconsin Press.

Cleveland, W. S. 1993. *Visualizing data.* Summit, NJ: Hobart Press.

———. 1994. *The elements of graphing data.* Summit, NJ: Hobart Press.

Cleveland, W. S., and R. McGill. 1984. Graphical perception: Theory, experimentation, and application to the development of graphical methods. *Journal of the American Statistical Association* 79: 531–554.

Conway, J. 1959. Class differences in general intelligence: II. *British Journal of Statistical Psychology* 12: 5–14.

Costigan-Eaves, P. 1984. Data graphics in the 20th century: A comparative and analytic survey. Ed.D. diss., Rutgers University.

Costigan-Eaves, P., and M. Macdonald-Ross. 1990. William Playfair (1759–1823). *Statistical Science* 5: 318–326.

———. (In progress). The method of curves: A brief history to the early nineteenth century. Open University, Milton Keynes, England. Typescript.

Daston, L., and P. Galison. 1992. The image of objectivity. *Representations* 40: 81–128.

De la Torre, L. 1952. *The heir of Douglas: Being a new solution to the old mystery of the Douglas cause.* New York: Knopf.

Donoho, A. W., D. L. Donoho, and M. Gasko. 1986. *MACSPIN: A tool for dynamic display of multivariate data.* Monterey, CA: Wadsworth & Brooks/Cole.

Dorfman, D. D. 1978. The Cyril Burt question: New findings. *Science* 201: 1177–1186.

Dynarski, M. 1987. The Scholastic Aptitude Test: Participation and performance. *Economics of Education Review* 6: 263–273.

Eastwood, B. 1987. Plinian astronomical diagrams in the early middle ages. In *Mathematics and its applications to science and natural philosophy in the middle ages,* ed. E. Grant and J. E. Murdoch, pp. 141–172. Cambridge: Cambridge University Press.

Edwards, D., and C. M. Beckworth. 1990. Comment on Holland and Wainer's "Sources of uncertainty often ignored in adjusting state mean SAT scores for differential participation rates: The rules of the game." *Applied Measurement in Education* 3: 369–376.

Eitelberg, M. J., J. H. Laurence, B. K. Waters, and L. S. Perelman. 1984. *Screening for service: Aptitude and education criteria for military entry.* Washington, DC: Office of the Assistant Secretary of Defense.

Farquhar, A. B., and H. Farquhar. 1891. *Economic and industrial delusions.* New York: G. P. Putnam's Sons.

Ferguson, Adam. *Britannica online.* http://www.eb.com:180/cgibin/g?DocF=micro/206/65.html.

Ferguson, S. 1991. The 1753 *Carte chronographique* of Jacques Barbeu-Dubourg. *Princeton University Library Chronicle* 52 (2): 190–230.

Fisher, R. A. 1936. The use of multiple measurements in taxonomic problems. *Annals of Eugenics* 7: 179–188.

Fitzgerald, P. H. 1904. *Lady Jean: The romance of the great Douglas cause.* London: Unwin.

Friedman, J. , J. W. Tukey, and P. A. Tukey. 1980. Approaches to analysis of data that concentrate near intermediate-dimensional manifolds. In *Data analysis and informatics,* ed. E. Diday, L. Lebart, J. T. Pagès, and R. Tomassone, pp. 3–13. Amsterdam: North Holland.

Friedman, M. 1992. Do old fallacies ever die? *Journal of Economic Literature* 30: 2129–2132.

Friendly, M. 2002. Visions and re-visions of Charles Joseph Minard. *Journal of Educational and Behavioral Statistics* 27 (1): 31–51.

Froncek, T. 1985. *An illustrated history of the city of Washington.* New York: Knopf.

Funkhouser, H. G. 1937. Historical development of the graphic representation of statistical data. *Osiris* 3: 269–404.

Funkhouser, H. G., and H. M. Walker. 1935. Playfair and his charts. *Economic History* 3: 103–109.

Galton, F. 1863a. A development of the theory of Cyclones. *Proceedings of the Royal Society* (London) 12: 385–386.

———. 1863b. *Meteorographica, or methods of mapping the weather.* London: Macmillan.

———. 1889. *Natural inheritance.* London: Macmillan.

Galton, S. T. 1813. *A chart, exhibiting the relation between the amount of Bank of England notes in circulation, the rate of foreign exchanges, and the prices of gold and silver bullion and of wheat accompanied with explanatory observations.* London: Published by J. Johnson & Co., St. Paul's Church-Yard; sold by J. Belcher and Son, Birmingham.

Gelman, A., and D. Nolan. 2002. *Teaching statistics: A bag of tricks.* Oxford: Oxford University Press.

Gilbert, E. W. 1958. Pioneer maps of health and disease in England. *Geographical Journal* 124: 172–183.

Graunt, J. 1662. *Natural and political observations on the Bills of mortality.* London: John Martyn and James Allestry.

Gunther, R. T. 1968. *Early science in Oxford*, vol. 13, *Dr. Plot and the correspondence of the Philosophical Society of Oxford.* London: Dawsons of Pall Mall.

Hambleton, R. K., and S. C. Slater. 1996. Are NAEP executive summary reports understandable to policy-makers and educators? Paper presented at the meeting of the National Council on Measurement in Education, New York, April.

Hearnshaw, L. S. 1979. *Cyril Burt, psychologist.* Ithaca, NY: Cornell University Press.

Herschel, J. F. W. 1833. On the investigations of the orbits of revolving double stars. *Memoirs of the Royal Astronomical Society* 5: 171–222.

Hilbert, D. 1902. Mathematical problems. *Bulletin of the American Mathematical Society* 8: 437–479.

Hoff, H. E., and L. A. Geddes. 1962. The beginnings of graphic recording. *Isis* 53: 287–324.

Hotelling, H. 1933. Review of *The triumph of mediocrity in business by H. Secrist. Journal of the American Statistical Association* 28: 463–465.

Huygens, C. 1895. *Oeuvres complètes*, vol. 6, *Correspondance.* The Hague: Martinus Nijhoff.

Jeon, J. W., H. Y. Chung, and J. S. Bae. 1987. Chances of Simpson's paradox. *Journal of the Korean Statistical Society* 16: 117–125.

Johnson, S. C. 1967. Hierarchical clustering schemes. *Psychometrika* 32: 241–254.

Kelley, T. L. 1927. *The interpretation of educational measurements.* New York: World Book.

King, W. I. 1934. Review of *The triumph of mediocrity in business* by H. Secrist. *Journal of Political Economy* 42: 398–400.

Kosslyn, S. M. 1994. *Elements of graph design.* New York: W. H. Freeman.

Larabee, L. W., ed. 1959–1988. *The papers of Benjamin Franklin*, vol. 15. New Haven, CT: Yale University Press.

Linn, R. 1982. Ability testing: Individual differences and differential prediction. In *Ability testing: Uses, consequences, and controversies*, ed. A. K. Wigdor and W. R. Garner, part 2, pp. 335–388. Report of the National Academy of Sciences Committee on Ability Testing. Washington, DC: National Academy Press.

Little, R. J. A., and D. B. Rubin. 1987. *Statistical analysis with missing data.* New York: Wiley.

Lombard, H. C. 1835. De l'influence des professions sur la durée de la vie. *Annales d'Hygiéne Publique et de Médecine Légale* 14: 88–131.

Madden, R. R. 1833. *The infirmities of genius, illustrated by referring the anom-*

alies in literary character to the habits and constitutional peculiarities of men of genius. London: Saunders and Otley.

Mangel, M., and F. J. Samaniego. 1984. Abraham Wald's work on aircraft survivability. *Journal of the American Statistical Association* 79: 259–267.

Marey, E. J. 1878. *La méthode graphique dans les sciences expérimentales et particuliérement en physiologie et en médecine*. Paris: n.p.

———. 1885. *La méthode graphique*. Paris: Boulevard Saint Germain et rue de l'Eperon.

Margerison, T. 1965. Review of Writing technical reports by Bruce M. Cooper. *Sunday Times* (London), 3 January.

Marsaglia, G. 1968. Random numbers fall mainly in the planes. *Proceedings of the National Academy of Sciences* 61: 25–28.

McKie, D. 1972. Scientific societies to the end of the eighteenth century. In *Natural philosophy through the 18th century and allied topics*, ed. A. Ferguson, pp. 133–143. London: Taylor & Francis.

Melvin, M. 1995. *International money and finance*. 4th edition. New York: Harper Collins College Publishers.

Meserole, M. 1992. *The 1993 information please sports almanac*. Boston: Houghton Mifflin Co.

Mills, F. C. 1924. *Statistical methods: Applied to economics and business*. New York: Henry Holt.

Mullis, I. V. S., J. A. Dossey, E. H. Owen, and G. W. Phillips. 1993. NAEP 1992: Mathematics report card for the nation and the states. Report no. 23-ST02, National Center for Education Statistics.

Newrick, P. G., E. Affie, and R. J. M. Corrall. 1990. Relationship between longevity and lifeline: A manual study of 100 patients. *Journal of the Royal Society of Medicine* 83: 499–501.

Page, E. B., and H. Feifs. 1985. SAT scores and American states: Seeking for useful meaning. *Journal of Educational Measurement* 22: 305–312.

Playfair, W. 1786. *The commercial and political atlas*. London: Corry.

———. 1801. *The commercial and political atlas*. 3d edition. London: John Stockdale.

———. 1805. *An inquiry into the permanent causes of the decline and fall of powerful and wealthy nations*. London: Greenland & Norris.

Plot, R. 1685. A letter from Dr. Robert Plot of Oxford to Dr. Martin Lister of the Royal Society concerning the use which may be made of the following history of the weather made by him at Oxford through out the year 1684. *Philosophical Transactions* 169: 930–931.

Powell, B., and L. C. Steelman. 1984. Variations in state SAT performance: Meaningful or misleading? *Harvard Educational Review* 54: 389–412.

Priestley, J. 1765. *A chart of biography*. London: William Eyres.

———. 1769. *A new chart of history*. London: n.p. Reprinted: New Haven, CT: Amos Doolittle, 1792.

Ramist, L., C. Lewis, and L. McCamley-Jenkins. 1994. *Student group differences in predicting college grades: Sex, language and ethnic group.* New York: College Board.

Randall, H. S. 1858. *The life of Thomas Jefferson.* New York: Derby and Jackson.

Randolph, L. F. 1934. Chromosome numbers in native American and introduced species and cultivated varieties of iris. *Bulletin of the American Iris Society* 52: 61–66.

Reilly, R. R. 1973. A note on minority group bias studies. *Psychological Bulletin* 80: 130–133.

Rosenbaum, P. R. 1989. Safety in caution. *Journal of Educational Statistics* 14: 169–173.

———. 1995. *Observational studies.* New York: Springer-Verlag.

Schmid, C. F. 1983. *Statistical graphics: Design principles and practices.* New York: Wiley.

Secrist, H. 1933. *The triumph of mediocrity in business.* Evanston, IL: Bureau of Business Research, Northwestern University.

Sharpe, W. F. 1985. *Investments.* 3d edition. Englewood Cliffs, NJ: Prentice-Hall.

Simpson, E. H. 1951. The interpretation of interaction in contingency tables. *Journal of the Royal Statistical Society* B 13: 238–241.

Smith, D. E. 1925. *History of mathematics*, vol. 2. Boston: Ginn & Co.

Smoking and health: Report of the Advisory Committee to the Surgeon General of the Public Health Service. 1964. Public Health Service Publication no. 1103. Washington, DC: Government Printing Office.

Sobel, D. 1999. *Galileo's daughter.* New York: Walker & Co.

Spence, I., and H. Wainer. 1997a. Who was Playfair? *Chance* 10 (1): 35–37.

———. 1997b. William Playfair: A daring worthless fellow. *Chance* 10 (1): 31–34.

———. 2001. William Playfair (1759–1823): An inventor and ardent advocate of statistical graphics. In *Statisticians of the centuries*, ed C. C. Heyde and E. Seneta, 105–110. New York: Springer-Verlag.

Steelman, L. C., and B. Powell. 1985. Appraising the implications of the SAT for educational policy. *Phi Delta Kappan* 67: 603–606.

Stigler, S. M. 1979. Psychological functions and regression effect. *Science* 206: 1430.

———1980. Stigler's law of eponymy. *Transactions of the New York Academy of Sciences* 2d series, 39: 147–157.

———. 1996. Adolphe Quetelet: Statistician, scientist, builder of intellectual institutions. A talk given at the Quetelet Bicentenary, Brussels, Belgium, October 24.

———. 1997. Regression toward the mean, historically considered. *Statistical Methods in Medical Research* 6: 103–114.

Taube, K. T., and K. W. Linden. 1989. State mean SAT score as a function of participation rate and other educational and demographic variables. *Applied Measurement in Education* 2: 143–159.

Thompson, D. W. 1961. *On growth and form.* Cambridge: Cambridge University Press.

Thoreau, H. D. 1968. *The writings of Henry David Thoreau: Journal,* vol. 2, *1850–September 15, 1851.* Ed. Bradford Torrey. New York: AMS Press. From the edition of 1906.

Thurber, J. 1939. *Fables for our time.* New York: Harper and Row.

Tilling, L. 1975. Early experimental graphs. *British Journal of Historical Science* 8: 193–213.

Tufte, E. R. 1983. *The visual display of quantitative information.* Cheshire, CT: Graphics Press.

———. 1996. *Visual explanations.* Cheshire, CT: Graphics Press.

———. 2000. Visual explanations. Yale University Tercentennial Lecture, November 15.

Tukey, J. W. 1939. *Convergence and uniformity in topology.* Annals of Mathematics Studies 2. Princeton, NJ: Princeton University Press.

———. 1962. The future of data analysis. *Annals of Mathematical Statistics* 33: 1–67, 812.

———. 1977. *Exploratory data analysis.* Reading, MA: Addison-Wesley.

———. 1986. Sunset salvo. *American Statistician* 40: 72–76.

———. 1989. Data-based graphics: Visual display in the decades to come. *Proceedings of the American Statistical Association* 84: 366–381.

U.S. Bureau of the Census. 1980. *Social indicators III.* Washington, DC: U.S. Bureau of the Census.

———. 1989. *Historical statistics of the United States: Colonial times to 1970.* White Plains, NY: Kraus International Publications.

U.S. Office of Management and Budget. 1973. *Social indicators,* 1973. Washington, DC: Government Printing Office.

Velleman, P. F. 1997. *DataDesk (version 6.0).* Ithaca, NY: Data Description Inc.

Wainer, H. 1980a. Making newspaper graphs fit to print. In *Processing of visible language* ed. P. Kolers, M. E. Wrolstad, and H. Bouma, vol. 2, pp. 125–142. New York: Plenum. Republished in 1981 in two parts: *Newspaper design notebook* (Chapel Hill, NC) 2 (6): 1, 10–16; and 3 (1): 3–5.

———. 1980b. A timely error. *Royal Statistical Society News and Notes* 7: 6.

———. 1984. How to display data badly. *American Statistician* 38: 137–147.

———. 1986. *Drawing inferences from self-selected samples.* New York: Springer-Verlag. 2d edition: Hillsdale, NJ: Lawrence Erlbaum Associates, 2000.

———. 1994. Three graphic memorials. *Chance* 7 (2): 52–55.

———. 1995. Graphical mysteries. *Chance* 8 (2): 52–56.

———. 1996. A Priestley view of international stock exchanges. *Chance* 9 (4): 31–33.

———. 1997a. Tom's veggies and the American way. *Chance* 10 (3): 40–42.

———. 1997b. *Visual revelations: Graphical tales of fate and deception from Napoleon Bonaparte to Ross Perot.* New York: Copernicus Books.

———. 2000a. Rescuing computerized testing by breaking Zipf's law. *Journal of Educational and Behavioral Statistics* 25: 203–224.

———. 2000b. *Visual revelations: Graphical tales of fate and deception from Napoleon Bonaparte to Ross Perot.* 2d edition. Hillsdale, NJ: Lawrence Erlbaum Associates.

Wainer, H., R. K. Hambleton, and K. Meara. 1999. Alternative displays for communicating NAEP results: A redesign and validity study. *Journal of Educational Measurement* 36: 301–335.

Wainer, H., C. Njue, and S. Palmer. 2000. Assessing time trends in sex differences in swimming and running. (With discussions.) *Chance* 13 (1): 10–15.

Wainer, H., S. Palmer, and E. T. Bradlow. 1998. A selection of selection anomalies. *Chance* 11 (2): 3–7.

Wainer, H., and D. Thissen. 1977. EXPAK: A FORTRAN IV program for exploratory data analysis. *Applied Psychological Measurement* 1: 49–50.

Wainer, H., and P. Velleman. 2001. Statistical graphics: Mapping the pathways of science. *Annual Review of Psychology* 52: 305–335.

Wald, A. 1980. A method of estimating plane vulnerability based on damage of survivors. CRC 432, July. (These are reprints of work done by Wald while a member of Columbia's Statistics Research Group during the period 1942–1945. Copies can be obtained from the Document Center, Center for Naval Analyses, 2000 N. Beauregard St., Alexandria, VA, 22311.)

Westbrooke, I. 1998. Simpson's paradox: An example in a New Zealand survey of jury composition. *Chance* 11 (2): 40–42.

Wilkinson, L. 1999. *The grammar of graphics*. New York: Springer-Verlag.

Woolsey, T. D. 1947. Adjusted death rates and other indices of mortality. In *Vital statistics rates in the United States, 1900–1940*, ed. F. E. Linder and R. D. Grove, pp. 60–91. Washington, DC: Government Printing Office.

Wurman, R. S. 1976. *What-if, could-be: An historic fable of the future.* Philadelphia: published by the author.

Yandell, B. H. 2002. *The honors class: Hilbert's problems and their solvers.* Natick, MA: A. K. Peters.

Yule, G. U. 1903. Notes on the theory of association of attributes of statistics. *Biometrics* 2: 121–134.

Zabell, S. 1976. Arbuthnot, Heberden, and the *Bills of mortality*. Technical Report no. 40. Department of Statistics, University of Chicago.

Zipf, G. K. 1949. *Human behavior and the principle of least effort.* Cambridge, MA: Addison-Wesley.

Index

Note: Page numbers followed by 'n' refer to a note on that page. *Italic numbers* refer to figures; **boldface numbers** refer to tables.